Innovations in Cognitive Neuroscience

Series editor

Vinoth Jagaroo
Department of Communication Sciences and Disorders
Emerson College
Boston, MA, USA

and

The Behavioral Neuroscience Program
Boston University School of Medicine
Boston, MA, USA

More information about this series at http://www.springer.com/series/8817

Maria Mody
Editor

Neural Mechanisms of Language

Springer

Editor
Maria Mody
Athinoula A. Martinos Center for Biomedical Imaging
Charlestown, MA, USA

Department of Radiology
Massachusetts General Hospital and Harvard Medical School
Boston, MA, USA

ISSN 2509-730X ISSN 2509-7318 (electronic)
Innovations in Cognitive Neuroscience
ISBN 978-1-4939-7323-1 ISBN 978-1-4939-7325-5 (eBook)
DOI 10.1007/978-1-4939-7325-5

Library of Congress Control Number: 2017953256

Printed on acid-free paper

This Springer imprint is published by Springer Nature
The registered company is Springer Science+Business Media, LLC
The registered company address is: 233 Spring Street, New York, NY 10013, U.S.A.

Note from the Series Editor

I am pleased to present this volume in the Springer Science + Business Publishers series *Innovations in Cognitive Neuroscience* which seeks to synthesize research and conceptual developments in cognitive science, neuroscience, and neurobiology. The accelerated pace of neuroscience research makes any topic in the field a moving target—and academic journals provide a fitting platform for such rapid developments. However, trends and trajectories in the field, often shaped by technological innovations and the new frameworks they present about functional models of brain and cognition, may be better captured and framed through the platform of a composed volume.

One of the stated themes of the series is the examination of cognitive domains such as memory, language, and spatial perception, from the perspectives of cutting-edge research and theory in behavioral neuroscience. It is within this segment of the series that the current volume on language is situated, yet the volume also relates well to other areas of the series centered on neuroanatomic and neural systems.

In the introduction to this concise volume, the editor Maria Mody summarizes a compelling selection of topics that help take us a step forward in our understanding of speech and language. The defining aspect of the volume is the advances in neuroscience offered though a new lens with which to examine human communication abilities. Specifically, advances in neuroimaging and associated analytic techniques have made for significant refinements in understanding structural-functional-connectivity patterns of the brain as they relate to language. Theoretical and mathematical models of neural network dynamics open the window into the links between oscillatory/frequency timescale patterns in neural networks and their sensori-cognitive correlates. And how these developments relate to speech and language, how they inform existing models or bring into question others, are well sampled by this volume. A variety of themes are covered, including acoustic processing of speech sounds, multisensory integration in language perception, semantic processing, supramodal theory, bilingual processing, cognitive vs. motor networks, neuroanatomic tracts and fasciculi, cerebellar-cortical systems in speech, the cortical functional architecture of language functions, and the subcortical basis of speech. It is, again, the appraisal of these themes sampled within a span of ten chapters, drawing on contemporary, cutting-edge technological, methodological, and analytic approaches, that defines the volume: What this means for some of the classical and canonical ideas about language, and what emerges in terms of new frameworks for the study of language, is elegantly conveyed by the volume.

The volume may be of particular interest to those in cognitive neuroscience and all related disciplines interested in the topic of brain and language. Researchers who employ functional imaging and electrophysiological investigations may also find interest in the methods and analytic models discussed.

Vinoth Jagaroo
Department of Communication
Sciences and Disorders
Emerson College
and
The Behavioral Neuroscience Program
Boston University School of Medicine
Boston, Massachusetts, USA

Contents

Contributors

Seppo P. Ahlfors Athinoula A. Martinos Center for Biomedical Imaging, Charlestown, MA, USA

Department of Radiology, Massachusetts General Hospital and Harvard Medical School, Boston, MA, USA

Michael C. Corballis School of Psychology, University of Auckland, Auckland, New Zealand

Marie-Christin Fellner Department of Neuropsychology, Institute of Cognitive Neuroscience, Ruhr University, Bochum, Germany

Elia Formisano Maastricht Brain Imaging Center, Maastricht University, Maastricht, The Netherlands

Karl J. Friston The Wellcome Trust Centre for Neuroimaging, University College London, London, UK

Simon Hanslmayr School of Psychology, University of Birmingham, Birmingham, UK

Niclas Kilian-Hütten Department of Psychiatry, Columbia University College of Physicians and Surgeons, New York, NY, USA

Department of Cognitive Neuroscience, Maastricht University, Maastricht, The Netherlands

Baojuan Li Athinoula A. Martinos Center for Biomedical Imaging, Charlestown, MA, USA

Department of Radiology, Massachusetts General Hospital and Harvard Medical School, Boston, MA, USA

School of Biomedical Engineering, Fourth Military Medical University, Xi'an, Shaanxi, China

Peter Mariën Department of Clinical and Experimental Neurolinguistics, Vrije Universiteit Brussel, Brussels, Belgium

Department of Neurology and Memory Clinic, Ziekenhuis Netwerk Antwerpen – Middelheim, Antwerp, Belgium

Katherine Midgley Department of Psychology, San Diego State University, San Diego, CA, USA

Maria Mody Athinoula A. Martinos Center for Biomedical Imaging, Charlestown, MA, USA

Department of Radiology, Massachusetts General Hospital and Harvard Medical School, Boston, MA, USA

Martin M. Monti Department of Psychology, UCLA, Los Angeles, CA, USA

Joanna Morris School of Cognitive Science, Hampshire College, Amherst, MA, USA

Dimitris Pinotsis The Picower Institute for Learning and Memory, Massachusetts Institute of Technology, Cambridge, MA, USA

Department of Brain and Cognitive Sciences, Massachusetts Institute of Technology, Cambridge, MA, USA

The Wellcome Trust Centre for Neuroimaging, University College London, London, UK

Sophie K. Scott Institute of Cognitive Neuroscience, University College London, London, UK

Jean Vroomen Department of Cognitive Neuropsychology, Tilburg University, Tilburg, The Netherlands

Chapter 1
Introduction: Speech and Language in the Brain—Framing Multiple Perspectives

Maria Mody

Over the last decade, there has been an unprecedented growth in our knowledge of the human brain, its structure and functioning, driven in large part by advances in neuroimaging tools and analysis methods. These developments have motivated researchers to push the boundaries of our understanding of human cognition, including language. This volume capitalizes on the advances, presenting multiple perspectives on language organization in the brain that take the reader from the seat of phonology and phonetics, to thought, meaning, and memory, and the evolutionary basis of our speech and language abilities. It combines various methods used to study language such as functional magnetic resonance imaging (fMRI), magnetoencephalography (MEG), and electroencephalography (EEG) with conceptual approaches incorporating multisensory perception, bilingual processing, effective connectivity, and causal modeling of interactions within the language network. Viewed through a wide-angle lens, the chapters provide a representative sampling of the breadth of language—its cognitive processes and neural subsystems—while also highlighting how advances in imaging/analysis techniques are driving the refinement of neurocognitive models of speech and language.

1.1 The Speech and Language Network

An increasing convergence of data from the neurosciences suggests that the brain is made up of a complex network of distinct neural circuits, aptly described as a hodotopic framework (Catani 2005). It includes cortical functional epicenters and

M. Mody (✉)
Athinoula A. Martinos Center for Biomedical Imaging, Charlestown, MA 02129, USA

Department of Radiology, Massachusetts General Hospital and Harvard Medical School, Boston, MA 02114, USA
e-mail: maria@nmr.mgh.harvard.edu

M. Mody (ed.), *Neural Mechanisms of Language*, Innovations in Cognitive Neuroscience, DOI 10.1007/978-1-4939-7325-5_1

connections between areas. As laid out by the now well-substantiated dual-stream model (Hickok and Poeppel 2004), the language network is made up of a dorsal stream associated with phonological processing via the superior longitudinal fasciculus (SLF), and a ventral stream associated with semantic processing supported mainly by the inferior fronto-occipital fasciculus (IFOF) in conjunction with the middle longitudinal fasciculus (MLF) and inferior longitudinal fasciculus (ILF). More recently, the frontal aslant tract (FAT) associated with initiating speech has been identified within the frontal lobe, connecting the medial aspect of the frontal lobe and the inferior frontal gyrus with the middle-inferior part of the pre-central gyrus (PCG; Catani et al. 2012). A meta-analysis of speech-related fMRI located the peak of activity within the laryngeal motor cortex in area 4p, the posterior part of the primary motor cortex, known to be involved in initiation and execution of motor commands (Simonyan 2014). Interestingly, damage to the anterior part of the neighboring PCG has been associated with a loss of prosody, whereas posterior damage leads to sound distortion errors (Tate et al. 2014).

Findings from graph-theoretical analysis, a mathematical technique widely applied in the understanding of neural network connectivity, have helped to further refine our understanding and delineation of the speech-specific network. Fuertinger et al. (2015) found that changes in network topology observed from fMRI resting state to syllable production (a speech-relevant motor task with minimal linguistic meaning) to speaking (complex production of meaningful sentences) when contrasted with the topology of a finger-tapping network (a non-speech-related motor control task) revealed a unique functional architecture for speaking: The speech-specific network consists of specialized cortical and subcortical nodes in the prefrontal cortex, insula, putamen, and thalamus. Additionally, the speech network requires only a specialized subset of the prefrontal region, suggesting that the prefrontal cortex may play a specialized role in the cognitive aspects of speech control such as verbal fluency, semantic context violations, and attention-demanding speech comprehension tasks.

A review of brain-based studies of speech and language reveals a variety of models with varying perspectives (Friederici et al. 2006; Hagoort 2005; Kuperberg et al. 2000; Rauschecker and Scott 2009). These are broadly represented in this volume—the chapters reflecting some of the differences between them originating in the neuroanatomy of speech perception, the dynamics of morphological processing, the functional architecture of semantics and memory, and the gestural origins of our capacity for speech and language, to name a few. Despite these differences, there is a promising amount of overlap between the various findings which provide useful insights about speech and language through different lenses. In the following sections we review the main points of the chapters within a framework of varying perspectives of speech and language.

1.2 Framework and Perspectives

At a basic level, spoken language processing entails encoding of the speech stream into its phonemic and syllabic constituents. However, this process is disrupted in many disorders of speech and language. Deficits in speech perception have been consistently associated with damage to the left superior temporal gyrus (STG) since the early observations of Carl Wernicke (1848–1905). The dorsolateral temporal lobe including the STG contains the primary auditory cortex and auditory association cortex, which explains why speech perception relies predominantly on auditory cortical fields. There is, however, considerable complexity in how speech and sound converge and diverge in acoustic processing. In Chap. 2, Scott points to relatively early sensitivity in the lateral STG for linguistically relevant aspects of sound over and above its acoustic structure. Particularly striking is the ability of the left STG to extract linguistic consistency from the constantly changing acoustics of the speech stream related to speaker and context effects.

An important consequence of a listener's sensitivity to acoustic–phonetic structure is the human ability to categorize speech sounds, paving the way for sound-symbol mapping essential for learning to read. However, as Morris points out (Chap. 3), whereas word recognition has a long history in psycholinguistics, few models of lexical processing include morphology as a level of linguistic representation, focusing mainly on morphologically simple words. It is morphologically complex words that help provide insight into how we recognize words and determine their meaning on the basis of the word's internal form. If complex words are decomposed, of interest are the neurophysiological markers of the process and the nature of the interaction among different levels of representation in the lexicon during recognition of complex words. Studies employing priming paradigms have found decision latencies to be shorter for targets preceded by primes with morphologic than orthographic relationships, and this has been found in many different languages. Neuroimaging findings show morphological processing in regions that also show orthographic priming like ventral temporal areas; that these regions do not overlap with those showing activation in semantic priming appears to support a morpho-orthographic account of lexical processing.

Unlike morphological processing, which relies on the internal form of words to determine their meaning, the semantic system may be viewed as providing a context for understanding words. Studies have found local and global neural assemblies to interact during semantic processing, mediated by synchronization and desynchronization of brain oscillations. In Chap. 4, Fellner and Hanslmayr draw on brain oscillations to provide special neurophysiological markers of language and memory. They focus on the inferior frontal gyrus (IFG) as a semantic hub involved in encoding, retrieval, and combining semantic content. Within this framework, the semantic system may be thought of as a distributed cortical network with increase in long-range gamma-band synchrony serving to bind information from different cortical areas for meaningful representation. In comparison, desynchronization in local neural assemblies, as in beta power decreases, is believed to support storage in memory.

These findings appear to reflect structural and processing hierarchies in speech and language mediated by synchronization and desynchronization of neural activity.

According to a dominant hypothesis in the field called the supramodal hierarchical parser (SHP) hypothesis, the center of language processing in the brain may be involved in processing hierarchical structures across domains of thought. The SHP implicates the left IFG in computations necessary for processing and representing abstract hierarchical "syntax-like" structured sequences across domains of human cognition. In Chap. 5, Monti explores this hypothesized relationship between mental computations that underlie our use of language and similar computations in other domains of human cognition, by examining neuroimaging and other data that converge on a functional interpretation of the left IFG. Perhaps, language serves as a founding mechanism providing the scaffolding on which capacities like logic and algebra have developed. Using a similar line of reasoning, the development of the semantic system may be seen as drawing on concepts already present in a prelinguistic system (Gelman and Gallistel 2004) with thought preceding language (Pinker 1984). Evidence, however, fails to support the SHP or that the left IFG and posterior temporal areas, which lie at the heart of language processing, are tuned to detect and represent complex hierarchical dependencies regardless of the specific domain of cognition.

1.3 Interacting with Language

Findings from connectivity studies show evidence of interactions between hierarchical sequences of sensory and cognitive activity within the language system, adding to the body of work of multimodal interactions in speech and language. Kilian-Hütten, Formisano, and Vroomen (Chap. 6) probe such interactions between streams of input (auditory and visual) highlighting the role of multisensory integration in speech and language processing. While it is well known that visual cues in the form of lip movements and gestures significantly enhance the perception and intelligibility of speech (Ross et al. 2006; Sumby and Pollack 1954), multisensory integration has been assumed to take place in higher order cortices. However, there is increasing evidence that integrative effects are already evident in the earliest stages of processing in unisensory cortical areas, with higher order regions involved in the convergence and integration of multisensory input as well as semantic and cognitive appraisal of this information. These findings are significant as they reveal cross-modal recalibration effects impacting later perception. They also raise an interesting question about bottom-up and top-down effects from competing languages on interactions within the language network in a bilingual speaker. For example, fluent access to two or more languages may influence the multisensory integration process at various points in the speech processing stream. In Chap. 7, Midgley focuses on the neurocognitive processing of words in one or both of bilinguals' two languages using EEG. Cross-linguistic data point to a single integrated lexicon in both languages for bilingual speakers, with lexical effects evident anywhere between 100 and 400 ms post-stimulus. The use of priming paradigms has been particularly effective in revealing interactions between levels of processing and isolating the loci of cross-linguistic effects (e.g., orthographic-semantic interface across different orthographic scripts). Words in a second language tend to recruit right-hemisphere visual word form areas in early processing (approximately 135 ms), and bilateral occipital temporal areas later on which disappears once fluency is achieved. This hints at plasticity differences in fast vs. slow second language learners. In summary, speech and language depends on an orchestration of multiple processes and structures in carefully synchronized patterns.

1.4 Looking Back in Time and in Location: Linking Structure and Function in the Brain

The last three chapters of this volume bring together various structural, functional, and evolutionary accounts of language in the brain. That the origin of speech may be traced to manual gestures highlights the ability to speak as reaching a level of autonomy only in the human species. The findings appear to support the view that the acquisition of speech as a higher order function depends on, and in turn is influenced by, sensorimotor and cognitive processes.

The structural and functional wiring of the brain provides a potential roadmap for probing these processes, the direction of information flow, and integration across levels within the speech and language network. In Chap. 8, Li, Ahlfors, Pinotsis, Friston and Mody review various measures used to investigate information flow between distributed regions within brain networks. They highlight the distinction between functional and effective connectivity, the latter providing explicit information about how distributed activity is caused and propagated within a network. The application of Granger Causality and Dynamic Causal Modeling (DCM) to speech and language data from young children and disordered populations has advanced the understanding of disorders of speech and language (Coben et al. 2014; David et al. 2011). These sophisticated analyses based on realistic neuronal models have helped identify perceptuomotor interactions within the speech network, and implicate both cortical and subcortical structures. As Mariën points out (Chap. 9), the cerebellum, which has been traditionally implicated in timing and rhythm in speech production, appears to be a crucial hub closely connected with frontal, temporal, limbic, and parietal association areas involved in higher cognitive and affective functioning. Studies with healthy subjects as well as patients with focal and diffuse cerebellar damage have confirmed the involvement of the cerebellum in word production as well as executive control via cerebello-cerebral connections to the prefrontal cortex. The application of causal modeling analyses to a distributed network incorporating these connections may help explain errors in normal speech (e.g., slips of the tongue) as well as impaired phonemic retrieval, shedding light on the cognitive-motor basis of speech production. Insofar as a variety of speech and language disorders have been found to follow cerebellar abnormalities, the findings lend support to the role of the cerebellum in higher cognitive functions, though there is still no consensus regarding the anatomical parts of the cerebellum that serve cognitive modulation.

That language may have evolved from manual gestures dovetails with various accounts of a role for the motor system in speech perception and speech production (Galantucci et al. 2006) and our earlier discussion of the cerebellum. Sign language in the deaf and correlation between handedness and asymmetry for language serve as additional evidence for the gestural origins of language. Furthermore, several studies have found evidence of motor deficits in developmental disorders, which appear to be associated with verbal deficits (Iverson and Fagan 2004; Gernsbacher et al. 2008; McCleery et al. 2013). The proposition of a subcortical basis for human language (Lieberman 2000) presents a significant challenge to the cortically based (evolutionary big-bang) notion of language evolution. Centered on this larger topic, Corballis, in the final chapter, brings us full circle—down the path of evolution and back to phonological gestures (Chap. 10)—by laying out evidence for the gestural origins of language.

Our understanding of the neural basis of language has clearly evolved from the early days of Broca and Wernicke. Rapid developments in the field of non-invasive neuroimaging, including diffusion tensor imaging (DTI) and transcranial magnetic stimulation (TMS) are yielding exquisite details about brain structure and connectivity. Methodological advances highlight the importance of functional and structural connectivity analyses in recognition of the brain working as a dynamic whole. A growing awareness of the role of frequency-specific neural oscillations in sensory and cognitive processing in combination with causal modeling is bringing us closer to more realistic models of speech and language processing grounded in the neurosciences.

References

Catani, M. (2005). The rises and falls of disconnection syndromes. *Brain, 128*, 2224–2239.

Catani, M., Dell'Acqua, F., Vergani, F., Malik, F., Hodge, H., Roy, P., et al. (2012). Short frontal lobe connections of the human brain. *Cortex, 48*, 273–291.

Coben, R., Mohammad-Rezazadeh, I., & Cannon, R. (2014). Using quantitative and analytic EEG methods in the understanding of connectivity in autism spectrum disorders: A theory of mixed over- and under-connectivity. *Frontiers in Human Neuroscience, 8*, 45.

David, O., Maess, B., Eckstein, K., & Friederici, A. (2011). Dynamic causal modeling of subcortical connectivity of language. *Journal of Neuroscience, 31*, 2712–2717.

Friederici, A., Bahlmann, J., Heim, S., Schubotz, R. I., & Anwander, A. (2006). The brain differentiates human and non-human grammars: Functional localization and structural connectivity. *Proceedings of the National Academy of Sciences, 103*, 2458–2463.

Fuertinger, S., Horwitz, B., & Simonyan, K. (2015). The functional connectome of speech control. *PLoS Biology, 13*, e1002209.

Galantucci, B., Fowler, C., & Turvey, M. (2006). The motor theory of speech perception reviewed. *Psychonomic Bulletin & Review, 13*, 361–377.

Gelman, R., & Gallistel, C. R. (2004). Language and the origin of numerical concepts. *Science, 306*, 441–443.

Gernsbacher, M., Sauer, E., Geye, H., Schweigert, E. K., & Hill Goldsmith, H. (2008). Infant and toddler oral- and manual-motor skills predict later speech fluency in autism. *Journal of Child Psychology and Psychiatry, 49*, 43–50.

Hagoort, P. (2005). On Broca, brain, and binding: A new framework. *Trends in Cognitive Sciences, 9*, 416–423.

Hickok, G., & Poeppel, D. (2004). Dorsal and ventral streams: A framework for understanding aspects of the functional anatomy of language. *Cognition, 92*, 67–99.

Iverson, J., & Fagan, M. (2004). Infant vocal-motor coordination: Precursor to the gesture-speech system? *Child Development, 75*, 1053–1066.

Kuperberg, G., McGuire, P., Bullmore, E., Brammer, M. J., Rabe-Hesketh, S., Wright, I. C., et al. (2000). Common and distinct neural substrates for pragmatic, semantic, and syntactic processing of spoken sentences: An fMRI study. *Journal of Cognitive Neuroscience, 12*, 321–341.

Lieberman, P. (2000). *Human language and our reptilian brain: The subcortical bases of speech, syntax, and thought*. Cambridge, MA: Harvard University Press.

McCleery, J., Elliott, N., Sampanis, D., & Stefanidou, C. (2013). Motor development and motor resonance difficulties in autism: Relevance to early intervention for language and communication skills. *Frontiers in Integrative Neuroscience, 7*, 30.

Pinker, S. (1984). *Language learnability and language development*. Cambridge, MA: Harvard University Press.

Rauschecker, J., & Scott, S. (2009). Maps and streams in the auditory cortex: Nonhuman primates illuminate human speech processing. *Nature Neuroscience, 12*, 718–724.

Ross, L., Saint-Amour, D., Leavitt, V., Javitt, D. C., & Foxe, J. J. (2006). Do you see what I am saying? Exploring visual enhancement of speech comprehension in noisy environments. *Cerebral Cortex, 17*, 1147–1153.

Simonyan, K. (2014). The laryngeal motor cortex: Its organization and connectivity. *Current Opinion in Neurobiology, 28*, 15–21.

Sumby, W., & Pollack, I. (1954). Visual contribution to speech intelligibility in noise. *The Journal of the Acoustical Society of America, 26*, 212–215.

Tate, M., Herbet, G., Moritz-Gasser, S., Tate, J. E., & Duffau, H. (2014). Probabilistic map of critical functional regions of the human cerebral cortex: Broca's area revisited. *Brain, 137*, 2773–2782.

Part I
Framework and Perspectives

Chapter 2
The Neural Processing of Phonetic Information: The Role of the Superior Temporal Gyrus

Sophie K. Scott

2.1 Introduction

Considerable advances have been made in our understanding of how the human brain processes the sounds of speech, since the earliest observations of (Wernicke 1874) that deficits in the perceptual processing of speech were associated with damage to the left superior temporal gyrus. As speech is a complex acoustic stimulus, Wernicke's observations were acute, since the dorsolateral temporal lobes (including the superior temporal gyri) contain primary auditory cortex and auditory association cortex. Though speech is the primary mode for linguistic expression in humans, its perception relies predominantly on auditory cortical fields. Subsequent work in the neuroanatomy of primate auditory cortex, in parallel with functional and anatomical studies of human auditory processing has indicated that there is considerable complexity in the ways that sound and speech is processed in the primate brain (Rauschecker 1998). In this chapter I will review some of the evidence for this from human functional imaging studies.

In the evolution of the brain, mammals are notable for their relatively large neocortex. Within the mammals, primates are notable for the relative diminution of the importance of their sense of smell (Moore 1981), and their (possibly corresponding) enhanced emphasis on the perceptual processing of visual and acoustic information (Felleman and Van Essen 1991). This is associated with increased complexity in the neuroanatomy of sensory processing in primates, with an anatomical hierarchy of the processing of sensory information in core belt and parabelt fields, areas that surround and are adjacent to the primary auditory cortex (PAC, A1). This anatomical hierarchy is mirrored by a hierarchy of function, where increased selectivity and/or complexity in cell responses to stimuli is associated with increasing distance from

S.K. Scott (✉)
Institute of Cognitive Neuroscience, University College London,
London WC1N 3AR, UK
e-mail: sophie.scott@ucl.ac.uk

M. Mody (ed.), *Neural Mechanisms of Language*, Innovations in Cognitive Neuroscience, DOI 10.1007/978-1-4939-7325-5_2

core fields. In visual cortex this has been associated with enhanced visual processing (Goodale 1998), for example, the enhanced color vision enjoyed by primates. In auditory processing there may be a similar association between enhanced neural systems and complex vocal behaviors in primates.

Core auditory fields (A1, rostral and rostro-temporal areas) in the primate brain receive all their input from the auditory thalamus (the medial geniculate nucleus/medial geniculate body, MGB). Whereas not all projections from the MGB go to core fields, the latter only receive input from the MGB (for complete reviews, look at Rauschecker 1998, Kaas and Hackett 2000). Core fields project to surrounding belt fields, and these in turn project to surrounding parabelt fields. There is a rostral-caudal organization to this relationship, with anterior core fields projecting to mid- and anterior belt fields, and posterior core fields projecting to mid- and posterior belt fields: in turn anterior belt fields project to mid- and anterior parabelt fields and posterior belt fields project to mid- and posterior parabelt fields. There is a tonotopy of responses in core fields, and this is preserved in cells moving laterally across the belt and parabelt. However, there is also an increasing sensitivity to the spectral complexity of sounds in a medial to lateral direction, with a gradient of responses revealing an increasing responsiveness to sounds with a widening spectral bandwidth. As we move further away from primary, core auditory cortex, there is an emerging pattern of segregation into at least two distinct anatomical pathways, associated with some specialization of function within the pathways. Thus in anterolateral auditory areas, there is a proportionally greater response to conspecific vocalizations (Tian et al. 2001), with some indication of a right hemisphere preference within this for aspects of caller identity (Petkov et al. 2008). In contrast, caudal auditory areas extending into the inferior parietal cortex are more sensitive to aspects of the spatial location of the sound (Tian et al. 2001), and some medial auditory areas show a response both to sound and to touch (Fu et al. 2003) (Fig. 2.1).

These studies have been tremendously important in helping us understand the neural processing of sound in the primate brain, but why is a chapter on phonetic processing starting with non-human primates, who may have many skills but who assuredly do not speak. Furthermore, spoken language was among the first cognitive functions to be localized, to some degree, in the human brain, due to the pioneering work of two neurosurgeons: Paul Broca's description of the importance of the posterior third of the left inferior frontal cortex to speech production (Broca 1865), and Carl Wernicke's demonstration of the involvement of the left superior temporal gyrus in receptive aphasia (Wernicke 1874). However, we focus on what bridges between these two fields, the developments of functional imaging which have shed a lot of light onto the neurobiology of speech perception over the last 15 years.

Functional imaging techniques such as positron emission tomography (PET) and functional magnetic resonance imaging (fMRI) which afford a relatively easy localization of activity in the human brain have two advantages which complement the classic neuropsychology approach. First they permit us to investigate speech and language in a healthy brain, rather than be constrained by the random ill fortune of neurological damage. Second they provide data which has meaning in a neuroana-

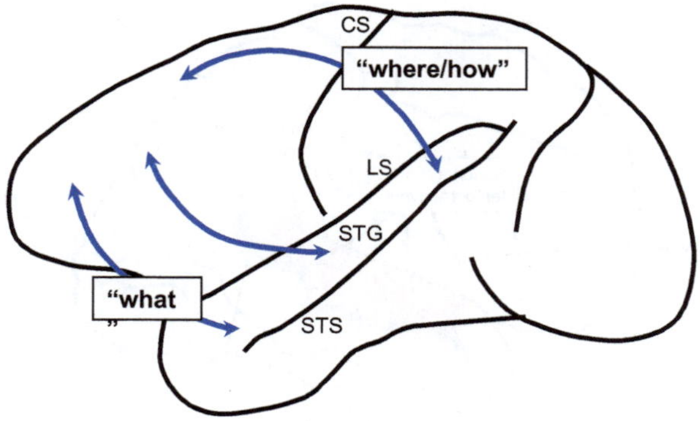

Fig. 2.1 Connections from auditory fields to prefrontal cortex in macaque monkeys. The blue arrows show the broad patterns of connectivity between anterior/rostral auditory areas and posterior/caudal auditory areas. These have been characterized as "what" and "where," and/or "how" pathways. Adapted from Kaas and Hackett (2000). *LS* lateral sulcus, *CS* central sulcus, *IPS* intra parietal sulcus, *STG* superior temporal gyrus, *STS* superior temporal sulcus

tomical frame of reference, which allows us to use theories and models from non-human neurophysiology and anatomy to interpret our results. The aim of this chapter is to use this neuroanatomical framework to explore the ways that phonetic information is processed in the human brain. Speech is a linguistic signal, but it is also an extremely complex sound so the structure and function of auditory cortex is an appropriate context in which to situate phonetic processing. In the light of this, the fact that sounds are processed in a relatively complex way in the primate brain—along different anatomical and functional streams—may have important implications for the neurobiology of speech perception. Not least, Wernicke's area as described by Wernicke (1874) to extend along the left superior temporal gyrus (Rauschecker and Scott 2009) may subtend a more complex range of auditory fields and acoustic processing potential.

2.2 Superior Temporal Gyrus in Humans

2.2.1 Structural Architecture

In humans, the dorsolateral temporal lobes are the site of auditory processing. Primary auditory cortex (A1) is found on the supratemporal plane, where Heschl's gyrus is a useful anatomical landmark. PAC is not necessarily precisely co-located with Heschl's gyrus but in most people it seems to be largely coextensive with the gyrus (also called the transverse temporal gyrus). The superior temporal gyrus (STG) is the cortical extent which is visible on the lateral aspect of the brain, extending laterally from within the Sylvian fissure, and which ends as the cortex folds into the superior temporal sulcus (STS) (Gloor 1997) (Fig. 2.2).

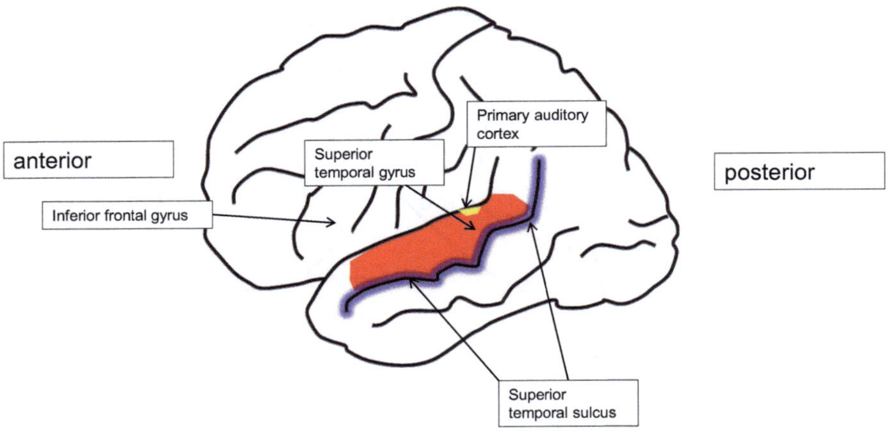

Fig. 2.2 Lateral surface of the human brain, showing the locations of primary auditory cortex, superior temporal gyrus (STG), superior temporal sulcus (STS), and inferior frontal gyrus (IFG)

Primary auditory cortex receives all of its input from the ascending auditory pathway, with specific projections from the auditory thalamus. In contrast, the majority of projections into auditory association cortex, which surrounds PAC (A1) and extends down to the STS, receive incoming projections from primary auditory cortex itself (though there are also sub-cortical projections to these fields; Pandya and Yeterian 1985). Work in humans has now revealed considerable complexity within PAC and auditory association cortex, with clear variation in the cytoarchitecture of different auditory fields around primary auditory cortex (Rivier and Clarke 1997; Wallace et al. 2002). Although we cannot do the kinds of single cell studies on humans which are possible in the non-human primate literature, so we cannot be precise about the relationship between a cell's functional responses and its anatomical properties, we can map between the results of functional imaging studies and these anatomical fields (Scott and Johnsrude 2003). Meta-analyses like these have indicated that primary auditory cortex does *not* show a speech specific response, unlike the predictions made by earlier speech theorists (Liberman and Whalen 2000) who identified that speech would be processed differentially from its earliest encoded entry into the cortex. Indeed, primary auditory cortex is sensitive to acoustic stimulation but is apparently showing a somewhat nuanced response beyond this: It shows no evidence for selectivity in its basic response, beyond some frequency specificity, but it does show context specific responses—cells in PAC will respond more strongly to infrequent sounds than when those same sound are presented frequently (Nelken and Bar-Yosef 2008; Ulanovsky et al. 2003). This may result from the more complex sub-cortical processing of sound than when compared to vision, for example, and may suggest that PAC is representing an interface between the representations of sound(s) in the ascending auditory pathway and higher-order cortical representations and processes. Notably, unlike ablation of visual cortex, which leads to cortical blindness and a lack of visual sensation, ablation of both auditory cortices does not result in deafness: sounds are still heard, but the perceptual properties of those sounds are harder to detect (Heffner and Heffner 1990). The functional differences in the cortical processing of visual and acoustic information may result from modality specific requirements; for example, sound is entirely temporal in nature and "glimpses" of sound are typically uninformative. The neural processing of sound may therefore have to be both faster than visual processing and also permit some buffering of the input such that sounds which are no longer available for perceptual processing are still available in a representational form for further elaboration (Scott 2005).

2.2.2 Functional Architecture

Areas in the lateral STG are sensitive to structure in sound, but are relatively heterogeneous in their response. FMRI studies have shown peaks in the lateral STG areas to harmonic structure (vs pure tones), amplitude modulations (vs unmodulated sounds), and to frequency modulations (vs sounds with a constant pitch) (Hall et al. 2002).

This lateral STG has also shown some sensitivity to speech relevant properties of sound. In an fMRI study, Davis and Johnsrude (2003) compared different neural activations associated with dimensions of speech intelligibility, using speech-in-noise at different signal-to-noise (SNR) ratios, noise vocoded speech involving different numbers of channels, and speech interrupted by noise at different interruption interval durations. This enabled them to identify neural responses which were sensitive to increasing intelligibility, and control for brain responses which were sensitive to the acoustic manipulations. When they did this, form independent intelligibility responses were seen in the superior temporal sulcus, whereas lateral STG areas showed a form dependent response, that is, a sensitivity to the acoustic properties of the sound in addition to any intelligibility-related responses. In contrast, however, studies which have used a more controlled approach to acoustic manipulations have identified a more complex profile of responses in the lateral STG area. In a study of phonotactic structure, Jacquemot et al. (2003) took non-word pairs that varied either phonetically or non-phonetically for both Japanese and French listeners: the non-words varied either in vowel duration or in syllable structure. Cleverly, they chose their phonetic/non-phonetic features as they were complimentary in Japanese and French. In Japanese, vowel duration is contrastive, so the non-words *ebuzu* and *ebuuzu* are phonetically distinct: to hear that they differ involves some phonetic processing. In French (as in English) vowel duration is not phonetically relevant, so to hear that *ebuzu* and *ebuuzu* are different involves more purely acoustic processing. In contrast, French permits (as does English) syllable structures which contain consonant clusters before the vowel (and after the vowel). Thus in French, *ebuzu* and *ebzu* are two different, phonotactically legal non-words. Japanese does not permit consonant clusters before (or after) the vowel, and Japanese speakers confronted with a non-word which contains such clusters (e.g., *ebzu*) will report hearing an "epenthetic" vowel between the consonants (e.g., *ebuzu*) (Dupoux et al. 1999). This means that if a French listener hears a difference between *ebuzu* and *ebzu*, then this involves some phonetic processing; if a Japanese listener hears a difference between *ebuzu* and *ebzu*, it is an acoustic difference, as both will be reported to sound like "*ebuzu*."

Jacquemot et al. (2003) took advantage of these differences between what constitutes a phonetic versus an acoustic change for Japanese and French listeners and used them in an fMRI study. Both French and Japanese listeners in the study were presented with pairs of non-words, which differed in either vowel duration or the presence of consonant clusters (or no difference). The contrast of greater activity for the phonetic condition (for both French and Japanese listeners) relative to the acoustic conditions was evident in the left lateral STG and the left SMG (Fig. 2.3).

The beauty of this study is that both language groups were tested with phonetic and acoustic differences which are complementary (what is a phonetic contrast for a French listener is an acoustic one for the Japanese and vice versa). The authors were thus able to associate the phonetic effects with linguistic, rather than purely acoustic properties of the stimuli. The study provided clear evidence for a relatively early sensitivity in the lateral STG for linguistically relevant aspects of sound, over and above its acoustic structure.

The result was confirmed in a study in which we varied the number of channels (viz., 1, 2, 3, 4, 8, 16) in noise vocoded speech (Scott et al. 2006). Noise vocoding is a technique in which the speech signal is broken down into a number of different channels, within which the amplitude modulation follows that in the corresponding speech signal. The more channels there are, the more the signal approaches the spectro-temporal profile of the original speech (Shannon et al. 1995). This manipulation gives a range of intelligibility scores which vary in a roughly logarithmic fashion with intelligibility. As increasing the number of channels in noise vocoded speech also increases the spectro-temporal complexity of the sentences, we included spectrally rotated sentences at 3 and 16 channels. This design allowed us to identify neural responses which showed the same shaped profile as the intelligibility responses, thus controlling for cortical responses which were similarly activated by the increase in acoustic complexity (Fig. 2.4).

This study revealed a network of bilateral STG/STS responses sensitive to different aspects of the sentences. Notably, the left lateral STG area was sensitive to the increasing number of channels, but was significantly more activated when these increases contributed to greater intelligibility in the speech, that is, the response to increasing number of channels was significantly greater to the speech than increasing the numbers of channels in rotated speech (which cannot be understood). It was not the case that the lateral STG was insensitive to the increasing number of channels in rotated noise vocoded speech—it was, but it was more activated when the same number of channels was present in a sentence which could be understood to some extent.

These findings suggest that, while the lateral STG is sensitive to acoustic structure, it is also sensitive to some aspects of linguistic information in the speech

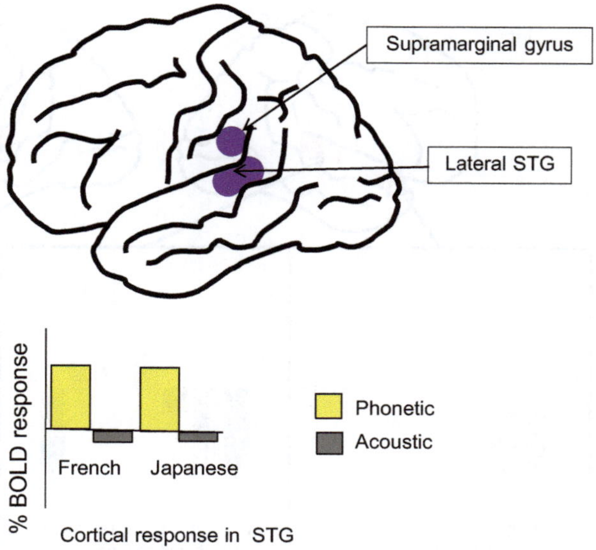

Fig. 2.3 Left lateralized cortical responses to phonetically relevant changes in stimuli, vs acoustical changes, across both Japanese and French listeners. The activations lie in the left lateral STG and supramarginal gyrus (adapted from Jacquemot et al. 2003). The lower panels show the response profiles for the STG peaks

signal—potentially at the level of the ways that phonetic information is expressed in the spectro-temporal variation in the speech signal. Notably, while research in vision has indicated considerable specificity of function in certain cell populations, auditory cortex has shown a much more heterogeneous response, both in single cell recordings (Eggermont 1998) and functional imaging studies (Brechmann and Scheich 2005). This flexibility may be strongly important for speech perception, as we know that individual phonetic contrasts can be cued by a host of independent acoustic factors (Lisker 1978), and that listeners must cope with variability in these both within and across speakers.

Consistent with this variability, studies from Obleser and colleagues have shown some evidence for abstract phonetic structure in the superior temporal gyrus (Obleser et al. 2006a). Using multiple exemplars of four stop consonants which varied in place of articulation and voicing, Obleser et al. revealed with magnetoencephalography (MEG) a topographic representation of phonetic properties of speech sounds (Obleser et al. 2006b). The analyses yielded an anterior–posterior source separation of the N100 response peaks to both alveolar stops (t, d) and the two velar stops (g, k). Notably, this difference was only significant for the speech tokens, and not for spectrally rotated baseline exemplars (which are unrecognizable) (Blesser 1972). The other clever aspect of this design was that there were many examples of each speech token, increasing the likelihood that the topological source separation was due to higher-order aspects of the speech sounds, and not to their acoustic variation (as there is a lot of acoustic variability within as well as between the speech conditions).

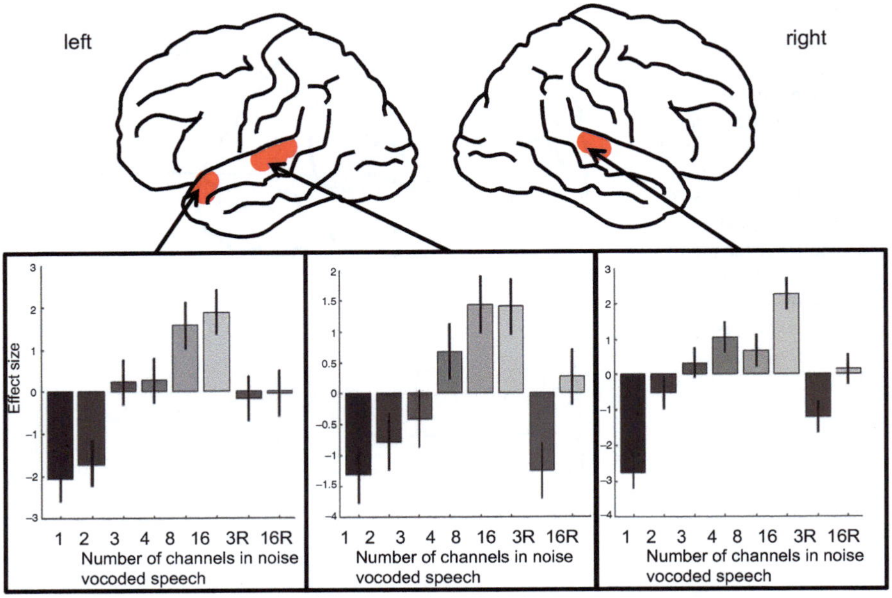

Fig. 2.4 Activations which correlate with intelligibility profiles for speech stimuli which have been noise vocoded to different numbers of channels, as well as for noise vocoded spectrally rotated speech (adapted from Scott et al. 2006). These are found in the *left* and *right* dorsolateral temporal lobes, and the *left anterior* temporal lobe. The lower panels show the response profiles for the three peak activations

2.3 Multiple Acoustic Sources

Another property of the plasticity and flexibility of acoustic processing in the STG is the capability to process multiple auditory streams. No sound is ever heard in silence, and there is a substantial literature on the ways that sounds are grouped into coherent streams/sources: this provides a way of thinking about the objects of auditory attention where at any one time, a single stream is being focused on in a way analogous to visual figure/ground grouping. The unattended streams are not discarded at the auditory periphery, and there is now considerable evidence that unattended sounds are processed to some degree, and that unattended sounds with a higher informational content—such as speech—are processed to a high level, such that words from an unattended stream can intrude into attention. This is referred to as informational masking in the psychoacoustic literature, and indicates some central competition for resources (Brungart 2001; Brungart and Simpson 2002).

In neural terms, one might expect this streaming to be associated with auditory cortex in the supra-temporal plane and STG. This has been borne out in a functional imaging study in which the number of concurrent sound sources, when varied, was positively correlated with activations in the anterior STG (Zatorre et al. 2002).

In contrast, a study of the perceptual control of the number of auditory streams perceived implicated the inferior parietal lobe, which may indicate that the manipulation of attention within candidate streams is not under purely auditory control (Cusack 2005). In terms of speech perception, several studies have indicated that speech is a strong source of informational masking. Functional imaging studies contrasting the perception of speech in the context of a speech masker, versus in continuous noise (commonly called energetic masking), have found extensive bilateral STG activation (Scott et al. 2004). This is consistent with the evidence for considerable processing of the unattended speech, and suggests that (a) STG can cope with at least two separate talkers' speech and (b) the neural processing of speech proceeds to some degree independent of attention mechanisms. Further studies have revealed that there are, along with attended speech, hemispheric asymmetries in this processing of unattended/masking speech: if unattended speech is intelligible, it leads to bilateral activation of the STG, but if it is not (e.g., through being spectrally rotated) this still leads to masking effects but leads to neural activation which is more right dominant (Scott et al. 2009).

There is clearly much more to know about these streaming/masking phenomena: we have addressed some of the issues around masking in a review (Scott and McGettigan 2013). However, the complexities of how these auditory streams are formed, how they interact and how attention interplays within them still need to be established in an auditory context. It is likely that speech perception mechanisms will involve cortico-fugal pathways between the ascending auditory pathways and the cortex (Colletti and Shannon 2005: Chandrasekaran et al. 2012).

2.4 Difference Between STG and STS

Functional imaging studies that contrast intelligible speech with an acoustic baseline typically reveal activation which runs along the superior temporal sulcus, ventral to the STG (Scott et al. 2000; McGettigan et al. 2011; Narain et al. 2003). The STS marks the end of auditory association cortex in primates, and the STS itself shows a heterogeneous response to different modalities, with a dominant response to social cues, for example, eye gaze, facial and body movements, and vocalizations (Allison et al. 2000). Responses are seen in the STS to speech that can be understood (Scott et al. 2000; Narain et al. 2003; Evans et al. 2014; McGettigan et al. 2011; Eisner et al. 2010) and also to CV syllables: unlike the STG, the STS is typically less sensitive to basic acoustic variation (although it has been shown to respond more to frequency modulated sounds than sounds with a constant frequency (Hall et al. 2002). It is possible therefore that activation in the STS reflects the outcome of acoustic–phonetic processing in the STG, and indeed some studies have overtly compared this response to an auditory "word form" area (Cohen et al. 2004).

The response in the STS to speech is typically left dominant, and with a greater response in the anterior (rostral) direction which has been overtly compared to the anterior "what" stream of processing in the primate brain (Scott and Johnsrude 2003;

Rauschecker and Scott 2009). This may reflect the entry point into the wider language system, including amodal semantic fields in the temporal pole(s) and the basal language area in the anterior fusiform gyrus (Patterson et al. 2007), as well as prefrontal areas (in medial and lateral prefrontal cortex), which are associated with the use of semantic context to support comprehension (Scott et al. 2003; Obleser et al. 2007).

2.5 Hemispheric Asymmetries

From Wernicke onwards, it is the left STG that has been directly implicated in phonetic perception. This is not an uncontroversial view, and theorists (Hickok and Poeppel 2000, 2004, 2007) have suggested that the pre-lexical processing of speech is a bilateral phenomenon, with both left and right STG fields contributing to the early acoustic–phonetic processing of speech, while functional imaging studies have indicated that right STG fields are strongly activated by speech (e.g., Mummery et al. 1999). However, this appears to be due to a right temporal lobe preference for a number of non-linguistic sources of information in the voice, from pitch variation (Scott et al. 2000; Johnsrude et al. 2000), to speaker identity (Scott et al. 2006; Belin and Zatorre 2003) and affective expressions (Schirmer and Kotz 2006). At the same time, the evidence for a left temporal lobe preference for linguistic information from acoustic phonetic structure (Jacquemot et al. 2003), through lexical (Cohen et al. 2004).semantic and syntactic processing (Friederici et al. 2003), is fully consistent with the clinical literature. As we find speakers easier to understand if we know them, and as we rapidly adapt to idiosyncrasies of speech production in a speaker specific fashion, these sources of information must interact and one of the greater problems in cognitive neuroscience is to understand these mechanisms.

The left dominant response to speech in auditory areas is consistent with the well-established role of the left hemisphere in language processing. There has been a historical interest in establishing a non-speech specific acoustic property that is preferentially processed on the left side of the brain, and which might lead to a left dominance via non-speech specific mechanisms. Candidate properties are rapid, or temporal processes (Zatorre and Belin 2001) and short time scale temporal structure (Poeppel 2003). Although these approaches have been immensely popular, they have been found wanting in terms of empirical evidence: it is now clear that the right temporal lobe does show some specific acoustic preferences, being activated more by sounds that are longer, or contain pitch variation (McGettigan and Scott 2012; McGettigan et al., 2012). Left auditory areas, in contrast, show a much more heterogeneous acoustic response, and only show a selective response when some linguistic properties are present in the signal—from phonetic structure through to semantic or syntactic detail. However, data from some recent studies involving passive listening to narrative speech show both hemispheres to be involved in speech perception (de Heer et al. 2017).

The question of whether this left temporal dominance in language processing can only be explained in linguistic terms remains open: just because a simple acoustic preference cannot explain the left dominance in speech perception (and production) does not mean that there could not be other candidate mechanisms. Notably, there is some evidence that expert perceptual processing may be associated with left temporal lobe mechanisms: a study with musicians who do and do not have perfect pitch showed a significantly greater activation of the anterior STS in musicians who have perfect pitch during a musical working memory task (Schulze et al. 2009). This may imply that expert listeners recruit this field to process sounds in which they are expert; however, since expertise in perceptual processing commonly co-occurs with an ability to categorize and name a stimulus (perfect pitch in terms of the ability to name a sound's pitch accurately), it may prove difficult to exclude linguistic mechanisms.

2.6 Phonemes in Superior Temporal Gyrus

The term phonetic processing is not synonymous with the claim that phonemes are perceptual constructs which have a neurobiological reality. In a recent functional imaging study we were able to show a significant STG activation to isolated unvoiced consonants (e.g., esh) relative to signal correlated noise equivalents, and also relative to ingressive speech sounds (e.g., a bilabial click) (Agnew et al. 2011). This result shows that the left STG is sensitive to phonetic structure, but not that phonemes necessarily form a stage of the perceptual processing of speech. Indeed, given the variation that is present in speech due to co-articulation, assimilation, and speaker variation, the auditory system is arguably better to look for patterns of covariation over sequences of speech sounds (Stilp et al. 2010). Consistent with this, in one study, participants were either directed to listen to non-words, or to silently rehearse them (McGettigan et al. 2011). The non-words varied in terms of number of syllable (2 vs 4) and the presence of consonant clusters (absent or present). In both the listen and the silent rehearsal groups, activation in the STG was strongly associated with the duration of the non-words (in terms of syllables). However, in neither listening nor rehearsing was the STG activation sensitive to the presence (or not) of consonant clusters. In contrast, left motor cortex was sensitive to both non-word duration and the presence of consonant clusters (Fig. 2.5).

While this is just one study, the evidence suggests that while STG may be sensitive to phonetic information, it may involve a stage of processing in which phonemes are selectively processed as distinct, discrete entities. The acoustic–phonetic profile of the sequence and the ways that it can be processed may be more important for the STG. In contrast, as the left dorsal precentral gyrus cortex is strongly implicated in the direct control of speech production, one might expect these fields to be more strongly driven by aspects of discrete phonetic identity, as these need to be individually implemented in speech production. Studies appear to suggest a close connection between brain systems for speech perception and production (Correia et al 2015), though the extent to which the brain's speech perception system uses articulatory representations during passive listening remains an open question.

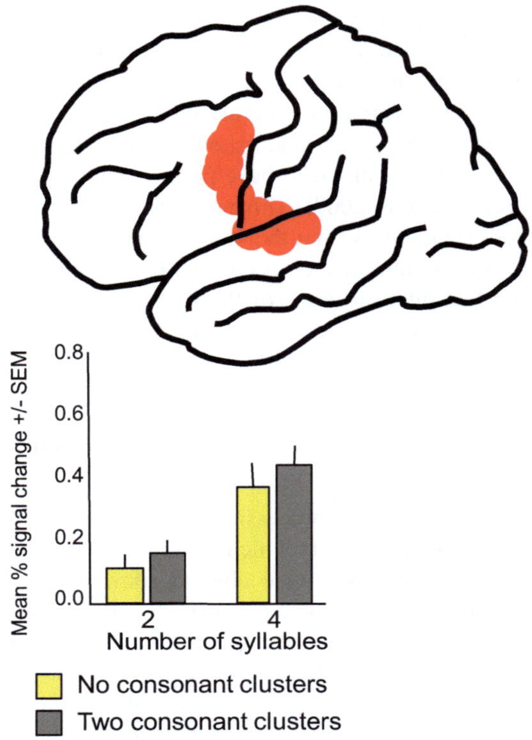

Fig. 2.5 Cortical responses which are sensitive to the number of syllables in non-words (2 or 4 syllables) during silent rehearsal (adapted from McGettigan et al. 2011). The lower panels show the response profiles for the peak response in the left dorsal precentral gyrus

2.7 Conclusions

In humans, the left superior temporal gyrus is critical to the early perceptual processing of phonetic information in speech. To be able to achieve this, processing in the left STG must be both fast, flexible and able to process multiple potential sources of relevant sounds. Processing in the STG is able to deal with great variety both across different talkers and within the same talker, as people's voice change greatly with context: the processing must be able to deal with the effects of noisy backgrounds, and also to process "unattended" auditory information to some degree, such that relevant information can compete for attention: it can cope in the presence of signal loss (e.g., when talking on the phone) and with novelty in the input—such as adapting to a new accent or a novel form of transformed speech. Developments in neuroscience have transformed our understanding of this area, but it is no exaggeration to say that we have really only just begun this journey.

References

Agnew, Z. K., McGettigan, C., & Scott, S. K. (2011). Discriminating between auditory and motor cortical responses to speech and nonspeech mouth sounds. *Journal of Cognitive Neuroscience, 23*, 4038–4047.

Allison, T., Puce, A., & McCarthy, G. (2000). Social perception from visual cues: Role of the STS region. *Trends in Cognitive Sciences, 4*, 267–278.

Belin, P., & Zatorre, R. J. (2003). Adaptation to speaker's voice in right anterior temporal lobe. *Neuroreport, 14*, 2105–2109.

Blesser, B. (1972). Speech perception under conditions of spectral transformation: I. Phonetic characteristics. *Journal of Speech, Language, and Hearing Research, 15*, 5–41.

Brechmann, A., & Scheich, H. (2005). Hemispheric shifts of sound representation in auditory cortex with conceptual listening. *Cerebral Cortex, 15*, 578–587.

Broca P (1865) Sur le siège de la faculté du langage articulé vol 6. V. Masson.

Brungart, D. S. (2001). Informational and energetic masking effects in the perception of two simultaneous talkers. *The Journal of the Acoustical Society of America, 109*, 1101–1109.

Brungart, D. S., & Simpson, B. D. (2002). Within-ear and across-ear interference in a cocktail-party listening task. *The Journal of the Acoustical Society of America, 112*, 2985–2995.

Chandrasekaran, B., Kraus, N., & Wong, P. C. (2012). Human inferior colliculus activity relates to individual differences in spoken language learning. *Journal of Neurophysiology, 107*, 1325–1336.

Cohen, L., Jobert, A., Le Bihan, D., & Dehaene, S. (2004). Distinct unimodal and multimodal regions for word processing in the left temporal cortex. *NeuroImage, 23*, 1256–1270.

Colletti, V., & Shannon, R. V. (2005). Open set speech perception with auditory brainstem implant? *The Laryngoscope, 115*, 1974–1978.

Correia, J.M., Jansma, B.M., & Bonte, M. (2015). Decoding articulatory features from fMRI responses in dorsal speech regions. *Journal of Neuroscience*, 35, 15015-15025.

Cusack, R. (2005). The intraparietal sulcus and perceptual organization. *Journal of Cognitive Neuroscience, 17*, 641–651.

Davis, M. H., & Johnsrude, I. S. (2003). Hierarchical processing in spoken language comprehension. *Journal of Neuroscience, 23*, 3423–3431.

de Heer, W.A., Huth, A.G., Griffiths, T.L., Gallant, J.L. & Theunissen, F.E. (2017). The Hierarchical Cortical Organization of Human Speech Processing. *Journal of Neuroscience*, 37, 6539-6557

Dupoux, E., Kakehi, K., Hirose, Y., Pallier, C., & Mehler, J. (1999). Epenthetic vowels in Japanese: A perceptual illusion? *Journal of Experimental Psychology, 25*, 1568.

Eggermont, J. J. (1998). Representation of spectral and temporal sound features in three cortical fields of the cat. Similarities outweigh differences. *Journal of Neurophysiology, 80*, 2743–2764.

Eisner, F., McGettigan, C., Faulkner, A., Rosen, S., & Scott, S. K. (2010). Inferior frontal gyrus activation predicts individual differences in perceptual learning of cochlear-implant simulations. *Journal of Neuroscience, 30*, 7179–7186.

Evans, S., Rosen, S., Golestani, N., Warren, J. E., McGettigan, C., Mourão-Miranda, J., et al. (2014). The pathways for intelligible speech: Multivariate and univariate perspectives. *Cerebral Cortex, 24*, 2350–2361.

Felleman, D. J., & Van Essen, D. C. (1991). Distributed hierarchical processing in the primate cerebral cortex. *Cerebral Cortex, 1*, 1–47.

Friederici, A. D., Rueschemeyer, S.-A., Hahne, A., & Fiebach, C. J. (2003). The role of left inferior frontal and superior temporal cortex in sentence comprehension: Localizing syntactic and semantic processes. *Cerebral Cortex, 13*, 170–177.

Fu, K.-M. G., Johnston, T. A., Shah, A. S., Arnold, L., Smiley, J., Hackett, T. A., et al. (2003). Auditory cortical neurons respond to somatosensory stimulation. *Journal of Neuroscience, 23*, 7510–7515.

Gloor, P. (1997). *The temporal lobe and limbic system*. New York: Oxford University Press.

Goodale, M. A. (1998). Vision for perception and vision for action in the primate brain. *Novartis Foundation Symposium, 218*, 21–34.

Hall, D. A., Johnsrude, I. S., Haggard, M. P., Palmer, A. R., Akeroyd, M. A., & Summerfield, A. Q. (2002). Spectral and temporal processing in human auditory cortex. *Cerebral Cortex, 12*, 140–149.

Heffner, H. E., & Heffner, R. S. (1990). Effect of bilateral auditory cortex lesions on sound localization in Japanese macaques. *Journal of Neurophysiology, 64*, 915-931.

Hickok, G., & Poeppel, D. (2000). Towards a functional neuroanatomy of speech perception. *Trends in Cognitive Sciences, 4*, 131–138.

Hickok, G., & Poeppel, D. (2004). Dorsal and ventral streams: A framework for understanding aspects of the functional anatomy of language. *Cognition, 92*, 67–99.

Hickok, G., & Poeppel, D. (2007). The cortical organization of speech processing. *Nature Reviews Neuroscience, 8*, 393–402.

Jacquemot, C., Pallier, C., LeBihan, D., Dehaene, S., & Dupoux, E. (2003). Phonological grammar shapes the auditory cortex: A functional magnetic resonance imaging study. *Journal of Neuroscience, 23*, 9541–9546.

Johnsrude, I. S., Penhune, V. B., & Zatorre, R. J. (2000). Functional specificity in the right human auditory cortex for perceiving pitch direction. *Brain, 123*, 155–163.

Kaas, J. H., & Hackett, T. A. (2000). Subdivisions of auditory cortex and processing streams in primates. *Proceedings of the National Academy of Sciences, 97*, 11793–11799.

Liberman, A. M., & Whalen, D. H. (2000). On the relation of speech to language. *Trends in Cognitive Sciences, 4*, 187–196.

Lisker, L. (1978). Rapid vs. rabid: A catalogue of acoustic features that may cue the distinction Haskins laboratories status report on speech. *Research, 54*, 127–132.

McGettigan, C., Evans, S., Rosen, S., Agnew, Z. K., Shah, P., & Scott, S. K. (2012). An application of univariate and multivariate approaches in FMRI to quantifying the hemispheric lateralization of acoustic and linguistic processes. *Journal of Cognitive Neuroscience, 24*, 636–652.

McGettigan, C., & Scott, S. K. (2012). Cortical asymmetries in speech perception: what's wrong, what's right and what's left? *Trends in Cognitive Sciences, 16*, 269–276.

McGettigan, C., Warren, J. E., Eisner, F., Marshall, C. R., Shanmugalingam, P., & Scott, S. K. (2011). Neural correlates of sublexical processing in phonological working memory. *Journal of Cognitive Neuroscience, 23*, 961–977.

Moore, W. J. (1981). *Mammalian skull*. Cambridge: Cambridge University Press.

Mummery, C. J., Ashburner, J., Scott, S. K., & Wise, R. J. (1999). Functional neuroimaging of speech perception in six normal and two aphasic subjects. *The Journal of the Acoustical Society of America, 106*, 449–457.

Narain, C., Scott, S. K., Wise, R. J., Rosen, S., Leff, A., Iversen, S., et al. (2003). Defining a left-lateralized response specific to intelligible speech using fMRI. *Cerebral Cortex, 13*, 1362–1368.

Nelken, I., & Bar-Yosef, O. (2008). Neurons and objects: The case of auditory cortex. *Frontiers in Neuroscience, 2*, 9.

Obleser, J., Boecker, H., Drzezga, A., Haslinger, B., Hennenlotter, A., Roettinger, M., et al. (2006a). Vowel sound extraction in anterior superior temporal cortex. *Human Brain Mapping, 27*, 562–571.

Obleser, J., Scott, S. K., & Eulitz, C. (2006b). Now you hear it, now you don't: Transient traces of consonants and their nonspeech analogues in the human brain. *Cerebral Cortex, 16*, 1069–1076.

Obleser, J., Wise, R. J., Dresner, M. A., & Scott, S. K. (2007). Functional integration across brain regions improves speech perception under adverse listening conditions. *Journal of Neuroscience, 27*, 2283–2289.

Patterson, K., Nestor, P. J., & Rogers, T. T. (2007). Where do you know what you know? The representation of semantic knowledge in the human brain. *Nature Reviews Neuroscience, 8*, 976–987.

Pandya, D. N., & Yeterian, E. H. (1985). Architecture and connections of cortical association areas. In Peters, A., & Jones, E. G. (eds.) *Cerebral Cortex, Volume 4, Association and auditory cortices* (pp. 3–61). New York: Plenum Publishing.

Petkov, C. I., Kayser, C., Steudel, T., Whittingstall, K., Augath, M., & Logothetis, N. K. (2008). A voice region in the monkey brain. *Nature Neuroscience, 11*, 367–374.

Poeppel, D. (2003). The analysis of speech in different temporal integration windows: Cerebral lateralization as 'asymmetric sampling in time'. *Speech Communication, 41*, 245–255.

Rauschecker, J. P. (1998). Cortical processing of complex sounds. *Current Opinion in Neurobiology, 8*, 516–521.

Rauschecker, J. P., & Scott, S. K. (2009). Maps and streams in the auditory cortex: Nonhuman primates illuminate human speech processing. *Nature Neuroscience, 12*, 718–724.

Rivier, F., & Clarke, S. (1997). Cytochrome oxidase, acetylcholinesterase, and NADPH-diaphorase staining in human supratemporal and insular cortex: Evidence for multiple auditory areas. *NeuroImage, 6*, 288–304.

Schirmer, A., & Kotz, S. A. (2006). Beyond the right hemisphere: Brain mechanisms mediating vocal emotional processing. *Trends in Cognitive Sciences, 10*, 24–30.

Schulze, K., Gaab, N., & Schlaug, G. (2009). Perceiving pitch absolutely: Comparing absolute and relative pitch possessors in a pitch memory task. *BMC Neuroscience, 10*, 106.

Scott, S. K. (2005). Auditory processing—Speech, space and auditory objects. *Current Opinion in Neurobiology, 15*, 197–201.

Scott, S. K., Blank, C. C., Rosen, S., & Wise, R. J. (2000). Identification of a pathway for intelligible speech in the left temporal lobe. *Brain, 123*, 2400–2406.

Scott, S. K., & Johnsrude, I. S. (2003). The neuroanatomical and functional organization of speech perception. *Trends in Neurosciences, 26*, 100–107.

Scott, S. K., Leff, A. P., & Wise, R. J. (2003). Going beyond the information given: A neural system supporting semantic interpretation. *NeuroImage, 19*, 870–876.

Scott, S. K., & McGettigan, C. (2013). The neural processing of masked speech. *Hearing Research, 303*, 58–66.

Scott, S. K., Rosen, S., Beaman, C. P., Davis, J. P., & Wise, R. J. (2009). The neural processing of masked speech: Evidence for different mechanisms in the left and right temporal lobes. *The Journal of the Acoustical Society of America, 125*, 1737–1743.

Scott, S. K., Rosen, S., Lang, H., & Wise, R. J. (2006). Neural correlates of intelligibility in speech investigated with noise vocoded speech—A positron emission tomography study. *The Journal of the Acoustical Society of America, 120*, 1075–1083.

Scott, S. K., Rosen, S., Wickham, L., & Wise, R. J. (2004). A positron emission tomography study of the neural basis of informational and energetic masking effects in speech perception. *The Journal of the Acoustical Society of America, 115*, 813–821.

Shannon, R. V., Zeng, F.-G., Kamath, V., Wygonski, J., & Ekelid, M. (1995). Speech recognition with primarily temporal cues. *Science, 270*, 303.

Stilp, C. E., Rogers, T. T., & Kluender, K. R. (2010). Rapid efficient coding of correlated complex acoustic properties. *Proceedings of the National Academy of Sciences, 107*, 21914–21919.

Tian, B., Reser, D., Durham, A., Kustov, A., & Rauschecker, J. P. (2001). Functional specialization in rhesus monkey auditory cortex. *Science, 292*, 290–293.

Ulanovsky, N., Las, L., & Nelken, I. (2003). Processing of low-probability sounds by cortical neurons. *Nature Neuroscience, 6*, 391–398.

Wallace, M. N., Johnston, P. W., & Palmer, A. R. (2002). Histochemical identification of cortical areas in the auditory region of the human brain. *Experimental Brain Research, 143*, 499–508.

Wernicke, C. (1874). The symptom complex of aphasia. In Cohen, R. S., & Wartofsky, M. W. (eds.) *Proceedings of the Boston colloquium for the philosophy of science 1966/1968, 1969* (pp. 34–97). Dordrecht, Holland: D. Reidel Publishing Company.

Zatorre, R. J., Bouffard, M., Ahad, P., & Belin, P. (2002). Where is 'where' in the human auditory cortex? *Nature Neuroscience, 5*, 905–909.

Zatorre, R. J., & Belin, P. (2001). Spectral and temporal processing in human auditory cortex. *Cerebral Cortex, 11*, 946–953.

Chapter 3
The Interaction Between Storage and Computation in Morphosyntactic Processing

Joanna Morris

To know a language means to be familiar with the units of representation that that language uses—phonemes, morphemes, words, and phrases—as well as with the rules by which those units combine so as to form meaningful utterances. The predictability of the relationship between form and meaning varies across linguistic forms. The meaning of a phrase is generally determined by the meaning of its constituent lexical forms, and the grammatical rules that specify how they combine. Nevertheless, there are idiosyncratic phrases that are semantically opaque and whose meaning cannot be derived transparently from their parts (e.g., "kick the bucket"). For simplex words or roots, the relationship between form and meaning is generally idiosyncratic. However, there are many complex words that transparently derive their meanings from their constituent parts, and thus can be recognized even when they have never been previously encountered (e.g., "to un-break"). Thus in both the morphological (word) and syntactic (phrase) domains, there are two systems that play a role in computing the relationship between form and meaning; an associative memory system—the mental lexicon—that stores arbitrary phonological, orthographic and conceptual–semantic relationships, and a system of rules—the grammar—that constrain the structure of complex forms.

In the following chapter, I will explore the relationship between these two systems by examining how they interact in the processing of morphologically complex words. Morphology has served as a fruitful arena of investigation for questions that speak to core issues of interest to all cognitive scientists, such as the trade-off in processing efficiency between computation and storage. Morphology is a combinatorial system in which simple units—roots and affixes—combine in systematic ways to produce complex words. The more information about complex words that is stored in memory, the fewer the online computational demands in comprehension and production; the less the information stored, the greater are the computational

J. Morris (✉)
School of Cognitive Science, Hampshire College, Amherst, MA 01002, USA
e-mail: jmorris@hampshire.edu

M. Mody (ed.), *Neural Mechanisms of Language*, Innovations in Cognitive Neuroscience, DOI 10.1007/978-1-4939-7325-5_3

demands. Is morphology explicitly represented in the language system, or does it emerge from the interactions between form and meaning representations? Are complex words decomposed and if so, under what circumstances? If decomposition does take place, how is this process instantiated in the brain and can we find neurophysiological markers of decomposition and recomposition? Does the recognition of complex words require two distinct kinds of processing mechanisms or can a single mechanism explain how all types of complex words are recognized? And finally, what is the nature of the interaction among different levels of representation in the lexicon during the recognition of complex words. In this chapter, I will first give definitions of the specific terminology used in the field. I will then describe the experimental paradigms that have typically been using the study morphological processing, and the models of morphological processing that have been proposed in the literature. Finally I will examine the behavioral, neuroimaging, and electrophysiological data that have been used to evaluate these models.

3.1 Deriving Meaning from Form

Recognizing a word involves determining what the word means on the basis of its form—its orthographic form in the case of reading, or phonological form in the case of listening. Words can be similar in meaning (synonyms) or form (homophones or homographs) but only morphologically related words share both meaning and form. Morphologically complex words can therefore play a key role in developing our understanding of the architecture of the processing system that maps formal orthographic and phonological representations onto semantic representations. Central to this architecture is the notion of the mental lexicon or mental "dictionary," the vast store that contains all our knowledge of the properties of individual words. (Aronoff 1983; Elman 2004).

Words in a language can either be simple, in which they cannot be further subdivided, or complex. If the latter, they can be further divided into meaningful subunits called morphemes. Morphemes are considered the smallest linguistic units that serve a grammatical function (Aronoff and Fudeman 2010). Lexical morphemes carry meaning in and of themselves, such as the words that generally fall into the syntactic classes of noun, verb, and adjective. In contrast, grammatical morphemes specify the grammatical relationships that hold between lexical morphemes, for example prepositions and determiners.

Morphemes can also vary in terms of their independence. Free morphemes can stand alone as words, but bound morphemes—generally affixes—cannot, and must be attached to either a root or stem. A stem is the part of the part of the word that is common to all its inflected variants, i.e., a base or root, plus any derivational affixes attached to it, e.g., "hunter" is the stem of the word "hunters." A root in contrast is the base form of a word or the form that exists after all affixes have been stripped away, e.g., "hunt" in "hunter." An affix is a bound morpheme that is added to either a root or stem to form a new word.

Affixes vary in the functions that they serve. Derivational affixes are used to derive new words from existing words, in the process, changing either the meaning and/or the syntactic category of the original. Inflectional affixes do not change the meaning or grammatical class of the words to which they attach, but rather modify a word's grammatical properties, such as verb tense or noun plurality. Affixes vary in terms of where they may appear with respect to the stem; they may appear before a stem (prefix), after a stem (suffix) within a stem (infix), surrounding a stem (circumfix) or between two stems (interfix). In English, all inflectional morphemes, for example, the past tense marker "-ed" and the plurality marker "-s" are suffixes. In contrast, derivational morphemes consist of both prefixes (e.g., re- and un-) and suffixes (e.g., –ment and –ness). Both infixes and circumfixes challenge the notion of a morpheme as a continuous discrete entity and present challenges for models of morphological processing that assume the existence of transparent segmentable morphemic units.

3.2 Methodological Considerations

3.2.1 Behavioral Methods

In order to understand how complex words are recognized we need to know both how morphological information is represented in the mental lexicon, and how it is stored and used during word recognition. Most researchers agree that morphological information becomes available during the processing of a complex word, but there is not yet consensus on if, or how, this information is represented in the mental lexicon, or the mechanisms via which this information is made available during word recognition.

Many of the studies that investigate morphological phenomena use the priming paradigm, in which a target word is preceded by a prime word that is morphologically related to the target word. Studies employing this paradigm have found that decision latencies to visually presented target words are generally shorter for targets preceded by primes with a morphological rather than a purely orthographic relationship although the morphological priming effect is typically less than that of identity priming (Fowler et al. 1985; Murrell and Morton 1974; Stanners et al. 1979). The morphological priming effect has been demonstrated in both immediate and long lag repetition priming (Bentin and Feldman 1990), in the auditory modality (Fowler et al. 1985; Kempley and Morton 1982), and in cross-modal priming (Feldman and Soltano 1999; Marslen-Wilson et al. 1994). These effects are not limited to English but have also been shown in Dutch (Drews and Zwitserlood 1995; Zwitserlood et al. 2005), German (Drews and Zwitserlood 1995), French (Meunier and Marslen-Wilson 2004; Meunier and Segui 1999), Italian (Laudanna et al. 1989; Orsolini and Marslen-Wilson 1997), Serbian (Feldman and Moskovljević 1987), Hebrew (Bentin and Feldman 1990; Bentin and Frost 1995; Feldman and Bentin 1994; Frost et al.

2000a, Frost et al. 1997) and Arabic (Boudelaa and Marslen-Wilson 2005, Boudelaa and Marslen-Wilson 2001). In addition, these studies have shown that morphological priming differs from purely orthographic and semantic priming in that morphological priming produces stronger and longer lasting facilitation than does semantic priming, while orthographic priming tends to result in inhibition (Drews and Zwitserlood 1995; Feldman 2000).

In order to dissociate priming effects mediated by orthographic or phonological representations from those mediated by more abstract modality-independent lexical representations, researchers have employed the cross-modal immediate repetition priming paradigm in which an auditory prime is immediately followed by a visual target. It is assumed that because the prime and target do not share any formal representations, any priming effects must be mediated by abstract modality-independent lexical representations and not via overlap in modality specific representations. However, Allen and Badecker (2002b) have argued that use the of the cross-modal paradigm "does not rule out the possibility that a spoken prime could influence (either through facilitation or inhibition) the form-level processing of a visual target at a lexical level. Nor does it guarantee, by separating the prime and target input modality alone, that no prelexical interactions between prime and target can arise." (p. 710).

The results of studies using visible primes can be difficult to interpret because it is unclear whether the observed priming effects are the result of automatic lexical processes (e.g., activation of a lexical entry) or the result of an episodic memory of the prime influencing the decision process to the later target. These studies are also vulnerable to the use of predictive strategies by participants if the relationship between prime-target pairs becomes obvious. These concerns can be partially addressed with the use of the masked priming paradigm (Forster and Davis 1984). In this paradigm, a prime is visually presented for a very brief period (approximately 50 ms). The prime is masked by the prior presentation of a masking stimulus, typically a series of hash marks (#####) or random consonant strings (SDFGHJK), and immediately followed either by another mask or by the target which serves as a backward mask. The short prime duration, as well as the presence of the forward and backward mask, prevents the subject from consciously perceiving the prime. Thus, any effects of the prime on responses to the target are presumed to reflect automatic persisting spreading activation from the prime, rather than strategic processes, or the effects of episodic memory (Grainger and Jacobs 1999; Van Heuven et al. 2001). If the prime and target are similar in form, then the target will reach the critical threshold for recognition more quickly (Forster 2009). As with visible primes, morphological priming effects using the masked priming paradigm have been consistently observed in the published literature (Forster et al. 1987; Frost et al. 1997; Giraudo and Grainger 2000, 2001; Grainger et al. 1991; Longtin et al. 2003; Meunier and Marslen-Wilson 2004; Pastizzo and Feldman 2004; Rastle et al. 2004; Voga and Grainger 2004).

Hypotheses about the representation and processing of complex words have been difficult to test using only reaction time paradigms as it is difficult to tease apart the influence of semantic, phonological, orthographic, and morphological and mnemonic properties on reaction time data. Moreover, reaction time measures are limited in that they reflect only the endpoint of the recognition process. The limitations of behavioral data have led researchers to supplement these data with other methodologies such as electroencephalography (EEG) involving scalp recorded event-related potentials (ERPs) and functional and structural magnetic resonance imaging (MRI). Below, we briefly recap some of these methods followed by a review of the findings using these methods that provide a close-up look at morphological processing in the brain.

3.2.2 Neurophysiological and Neuroanatomical Methods

EEG

ERPs are voltage changes in the ongoing electroencephalogram that are time-locked to the onset of a sensory or motor event (Garnsey 1993; Luck 2014). ERPs allow us to closely tie cognitive processes to brain function. If it is assumed that distinct processes are mediated by different underlying neurophysiological and neuroanatomical mechanisms, differences in ERP duration, amplitude, polarity, and scalp topography can provide evidence for distinct brain, and by extension cognitive, mechanisms (Osterhout 1997). ERPs are well suited to the study of language processing because they allow for the tracking of perceptual and cognitive processes as they unfold, without requiring participants to produce overt responses that may interfere with the cognitive events related to stimulus processing. Electrophysiological techniques are particularly well-suited to examine potential early effects of morphological structure as they reflect neural processing on a continuous millisecond by millisecond basis and thus have the temporal resolution needed to capture perceptual and cognitive processes that occur well before the behavioral response.

Grainger and Holcomb have proposed a mapping of the ERP components observed in their masked repetition priming experiments onto component processes in a functional architecture for visual and spoken word recognition, the bi-modal interactive activation model (BIAM). These components—including the N250 and N400—whose amplitudes are modulated by priming, appear to reflect processing that proceeds from visual features to orthographic representations and finally to meaning (Grainger and Holcomb 2009, Grainger et al. 2010; Holcomb and Grainger 2006).

The N250 is a negative-going wave that starts as early as 110 ms and peaks around 250 ms. It is sensitive to the degree of prime-target orthographic overlap, being larger for targets that share only partial orthographic overlap with their primes compared to targets that share complete overlap (Holcomb and Grainger 2006). Holcomb and Grainger suggest that the N250 reflects the process that maps sublexical orthographic and phonological representations onto whole word representations. Within the general framework of interactive-activation, this mapping process involves the transfer of activation (feed-forward and feed-back) from letter and letter cluster representations to whole-word representations, and therefore indexes the initial build-up of activation at the lexical level and the stabilization of activation at the sub-lexical level. Moreover, the N250 may not be a unitary entity but rather may be composed of distinct sub-components that index separable orthographic and phonological processes in word recognition. In a priming study combining an orthographic (transposed letter), and phonological (pseudohomophone), manipulation of the prime, Grainger et al. (2006) found a distinction between an early posterior N250, driven by the transposed letter primes, and a late anterior N250 driven by the pseudohomophone primes. These data suggest that the early N250 effect may reflect pre-lexical orthographic processing, and the late effect pre-lexical phonological processing. The apparent sensitivity of the N250 to pre-lexical processing suggests that it can be productively used to investigate the early cognitive processes involved in complex word recognition.

The N400 is a negative going component with a central-posterior maximum which Holcomb and Grainger (2006) interpret as reflecting the amount of effort involved in forming links between word and concept representations with larger N400s indicating a more effortful process. The N400 (at least in its early phase, the N400w) is hypothesized to reflect the mapping of lexical form onto meaning while later effects could reflect integration across semantic representations. Translated into the mechanics of interactive-activation, this mapping process reflects the transfer of activation from whole-word representations to semantic representations and therefore indexes the initial phase of activation of semantic representations and the stabilization of activation at the level of whole-word units.

MEG and fMRI

Magnetoencephalography (MEG) is a functional neuroimaging technique for mapping brain activity by recording magnetic fields produced by the electrical activity of the brain. The MEG component of most utility in elucidating the time course and neuroanatomical boundaries of early morphological processing is the M170 (the MEG counterpart to the N170 evoked by visually presented objects). The M170 has generators in ventral temporal lobe and shows sensitivity to the frequency properties of letter strings and appears to corresponds to the fMRI response from the so-called visual word-form area (Solomyak and Marantz 2010).

Functional neuroimaging methods such as functional magnetic resonance imaging (fMRI) and positron emission tomography (PET) are based on the premise that changes in cerebral activity during the performance of a task are associated with changes in regional cerebral blood flow. Functional neuroimaging techniques have superior spatial resolution and yield precise mappings of brain areas involved in task performance, but have much poorer temporal resolution than electro- and magnetophysiological techniques.

3.3 Models of Morphological Processing

Morphology has served as a fruitful arena for debating the closely linked questions about the psychological status of grammatical rules, and the relative strengths of connectionist and symbolic models of cognition (Pinker 1991, Pinker 1997; Pinker and Ullman 2002). Behavioral effects of morphological structure are generally taken to lend support to models of word processing in which morphology is explicitly represented, either because complex words are stored and access via their constituent morphemes or because morphologically related words share connections in the lexicon. In contrast to symbolic models with explicit representations of morphology, connectionist models attempt to account for the behavioral effects of morphological structure without an explicit representation of morphology in the lexicon. These effects are accounted for by shared patterns of activation across semantic, orthographic, and phonological processing units (Raveh 2002; Rueckl et al. 1997; Rueckl and Raveh 1999; Seidenberg and Gonnerman 2000).

3.3.1 Single Mechanism Models

The morphological processes of inflection and derivation alter the form of a word—by the addition of an affix, or by changing the form of a root—in order to indicate grammatical features such as number, person, mood, or tense, or to derive another word differing in meaning or syntactic category. Given that morphological processes generally consist of systematic changes to the form of a stem resulting in predictable changes to that stem in meaning, many early models of morphological processing posited that complex forms are generated by rules. Rule-based processes work well for complex forms that are transparently compositional such as regular inflection (walk ⇨ walked) and semantically transparent derivation (good ⇨ goodness) but are less well-suited to explain irregular inflection (teach ⇨ taught) or semantically opaque derivation (apart ⇨ apartment) in which the output of the process is unpredictable.

There are, however, models of morphological processing that attempt to extend a rule-based approach to the processing of irregular forms (Chomsky and Halle 1968; Halle and Marantz 1994; Stockall and Marantz 2006). These models posit "full, across the board, decomposition" (Stockall and Marantz 2006), with no categorical differences between regular and irregular allomorphy, or between concatenative (affixal) and non-concatenative morphology, such as the stem changes that are common to irregular past tense formation in English. These models use slightly different formalisms to account for the variation in the realization of the past tense morpheme and the stem allomorphy across irregular roots, but all claim that processing the irregular inflected form (e.g., "taught") involves activating the root (e.g., "teach") by virtue of successfully recognizing the surface phonological string

("-aught") as the output of a rule that operates over the underlying phonemic sequence comprising the root morpheme.

In contrast to models that posit that all complex forms are generated by rule, there are two types of models that propose that all previously encountered inflected words, both regular and irregular, are stored as whole forms in memory. In connectionist models, all inflected forms, whether familiar or new, are represented in a distributed associative memory (Plunkett and Juola 1999; Plunkett and Marchman 1993; Rumelhart and McClelland 1986). These models assume a single pattern associator network in which word recognition proceeds via the establishment of stable activation states over sets of input and output units that represent the orthographic, phonological, and semantic properties of a word. Morphological effects are accounted for by shared patterns of activation across semantic, orthographic, and phonological processing units. Thus connectionist models do not make use of symbolic rules for regular inflection, but treat all inflection as a set of mappings from the root to the past tense form.

In full listing models all words are stored in the lexicon regardless of their morphological complexity (Butterworth 1983; Manelis and Tharp 1977; Rubin et al. 1979). In such models, affixed words are processed in their entirety, without the need for decomposition. Full-listing models were proposed to account for data showing no differences in response times between truly affixed and pseudo-affixed words (Manelis and Tharp 1977; (Rubin, Becker, & Freeman, 1979) Thus the debate among models of morphological processing has focused on models that include mechanisms for representing morphological relationships, both implicitly and explicitly.

3.3.2 Dual-Mechanism Models

Full-parsing models are ones in which all morphologically complex words that are transparently segmentable (i.e., with separable stems and affixes) are recognized on the basis of their constituents. Taft proposed an "affix-stripping" model (Taft 1979b, 1994; Taft and Forster 1975) in which prefixed words are represented by their stem in a peripheral orthographic access lexicon, and these stems act as modality specific codes through which information about the full word forms in a central lexicon is made available. Later formulations of the model adopt an interactive activation framework in which morpheme units intervene between the grapheme and whole word representations (see Fig. 3.1).

Full-parsing models are problematic, however, in cases where there are several alternative parses for a given input, when there are pseudo-affixes or pseudo-stems, and for opaque word forms, whose meaning cannot be derived from the meaning of their constituent morphemes.

Dual-route models (Caramazza et al. 1988; Diependaele et al. 2012; Frauenfelder and Schreuder 1992; Pinker 1991; Schreuder and Baayen 1994) combine full-listing and full-parsing models by proposing two possible pathways to recognition; a direct route in which complex words are recognized on the basis of their surface form, and

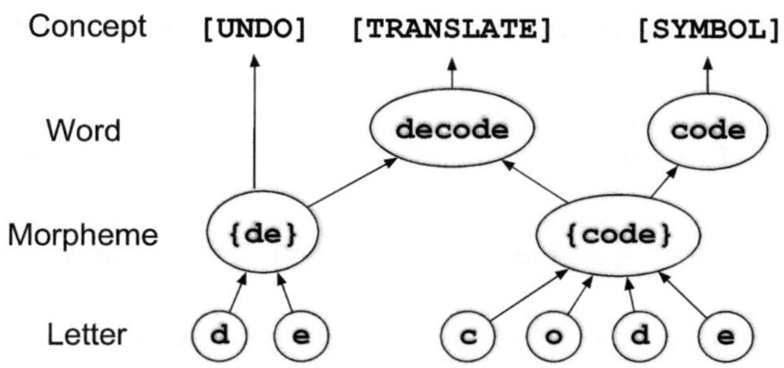

Fig. 3.1 Taft's interactive activation model in which stems and prefixes are treated separately and constitute activation units

a decompositional route in which these words are broken down into their constituent morphemes and recognized via their stems.

The Augmented Addressed Morphology (AAM) model (Burani and Caramazza 1987; Caramazza et al. 1988; Laudanna and Burani 1995) posits that familiar complex words are recognized via the direct route, while rare or unknown words that are transparently compositional are recognized via the decompositional route. In the dual-route Morphological Race (MR) model of Schreuder, Baayen and colleagues, (Baayen et al. 1997; Frauenfelder and Schreuder 1992; Schreuder and Baayen 1995), as in the AAM model, the decompositional and the direct full-forms routes operate in parallel. In the MR model when a morphologically complex word is processed, the direct route undertakes a search of the lexicon for the surface form of the word. At the same time, the decompositional route breaks the word down into its constituent stems and affixes. The two routes compete to recognize the input with the outcome determined by the faster route. The speed with which the parser is able to compute a meaning for the full form is a function of both the frequency of the constituent stems and affixes, and the semantic transparency of their combination.

The Words and Rules (WR) model of Pinker and colleagues (Pinker 1991, 1997; Pinker and Ullman 2002) is a dual-route model designed to account for the production of regular inflections and thus has little to say about the processing of derivations. Nonetheless, to the extent that derivations are predictable in form and semantically transparent, they should be processed in a manner similar to regular inflections. According to this model, regular inflection is computed by a rule that concatenates a specified affix to a stem of a particular category. Irregular forms, in contrast, are unpredictable and so cannot be produced by a rule but must be stored in, and retrieved from, memory. This model is also in a sense a race model in that irregular forms are retrieved quickly via a direct route to memory, and the retrieval of an irregular forms blocks application of the regular rule. However, when retrieval fails, the combinatorial route applies and the regularization of an irregular form results.

In dual-route models that propose (a) the existence of both full-form and sub-lexical morphemic representations, as well as (b) a hierarchically organized

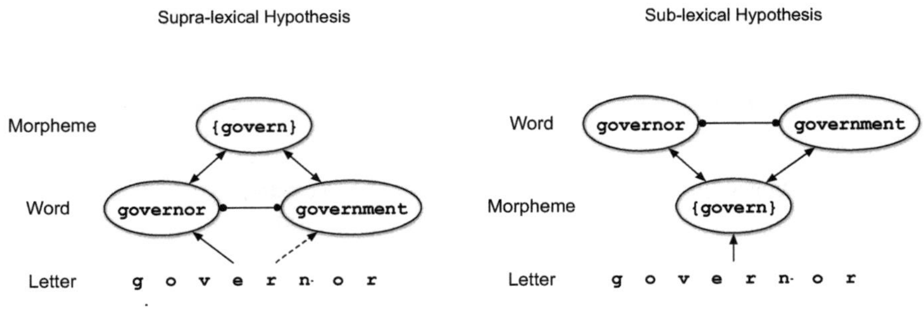

Fig. 3.2 Supra-lexical and sub-lexical morphemic representations in a hierarchically organized morphological processing framework. In the sub-lexical model morphemic representations are form-based (orthographic and phonological); in the supra-lexical model they are modality independent

processing stream in which lower level form representations are activated before higher level semantic ones, morphemic representations may be located above (supra-lexical) or below (sub-lexical) the level of whole-word orthographic or phonological representations. Sub-lexical morphemic representations are purely formal, whereas supra-lexical representations are amodal and form an interface between form-based lexical representations on the one hand and semantic representations on the other. Giraudo and Grainger (2000, 2001) proposed a supra-lexical model of morphological processing in which complex words are represented as whole units at the level of form, but morphological relatives are linked to common morphological representations at a higher level of linguistic structure, with the patterns of connectivity determined by semantic transparency. Thus in this model, morphological relatedness is determined only after lexical access has occurred. This model predicts that only semantically related morphological primes will facilitate target processing relative to unrelated primes (see Fig. 3.2).

However, the finding of robust masked morphological priming effects for semantically opaque words (e.g., apartment-apart, corner-corn) suggests that there is a form of early morphological decomposition that is based on orthographic rather than semantic information. Thus, reading complex words may involve a rapid morphemic segmentation based purely on the analysis of orthography, which has been coined "morpho-orthographic decomposition" (Rastle and Davis 2008; Rastle et al. 2004).

An alternative proposal, put forth by Diependaele, Grainger and colleagues (Diependaele et al. 2012; Diependaele et al. 2013; Diependaele et al. 2005, Diependaele et al. 2009; Morris et al. 2011), includes a role for both orthographic and semantic representations in the early stages of word recognition. This proposal suggests that the internal morphological structure of printed words is represented at both the sub- and supra-lexical levels. In this model, representations at the lexical level can be activated via sub-lexical morpho-orthographic representations, but they can also be activated by a route that attempts to map the whole input directly onto a corresponding whole-word form representation. Lexical representations are in turn

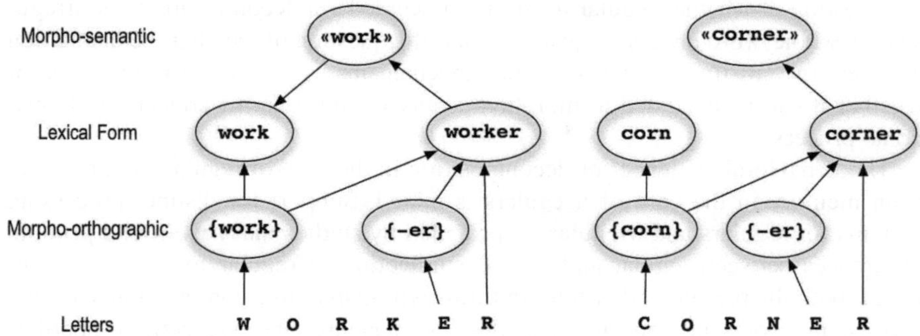

Fig. 3.3 Dual-route model of morphological processing with both morpho-orthographic and morpho-semantic processing routes

mapped onto higher-level morpho-semantic representations that code for the similarity in form and meaning between known words, i.e., morphological relatedness (Giraudo and Grainger 2000, 2001; Grainger et al. 1991). In a priming context, opaque morphological relatives will only be able to prime each other through shared representations at the morpho-orthographic level, whereas transparent items will also be able to do so via the shared representations at the morpho-semantic level (see Fig. 3.3). Therefore, what distinguishes the morpho-orthographic and morpho-semantic dual-route accounts is the role of meaning in early morphological decomposition, a distinction that has deep implications for the construction of models of the skilled reading process. In the following section, I will examine the evidence for and against each of these models, by examining how the process of word recognition can be modulated by morphological regularity, lexical frequency, and semantic transparency.

3.4 Morphological Processing Effects

3.4.1 Regularity Effects

Regularity may be one dimension along which the distinction between decomposition and storage is drawn. The Word and Rules model of Pinker and colleagues (Pinker 1991, 1997; Pinker and Ullman 2002; Prasada and Pinker 1993) posits that regular inflection proceeds via abstract symbolic rules that concatenates a suffix to a symbol for a syntactic category. The complex words produced in this way are semantically transparent and fully decompositional in terms of their form, such that the meaning of the complex form can be determined solely on the basis of the meanings of its constituent morphemes. Regularity is generally considered in the context of inflectional morphology, but in some cases derivation may be as regular as and as productive as inflection (Clahsen 2003; Raveh and Rueckl 2000). Thus, alongside the question of whether regular forms are processed via decomposition and irregulars via whole-word processing, we can raise the question of whether the distinction between two distinct representational mechanisms is specific to inflection or whether it can be extended to morphologically complex forms created by derivational processes.

The behavioral evidence for decomposition in the case of regulars and retrieval from memory in the case of irregulars is mixed. Support for distinct processing systems for regulars and irregulars is provided by studies that have found priming differences between regular and irregular inflection. If regular forms are decomposed, both the present and past tense forms will map onto a common entry for the verb stem representation that encodes its meaning and syntactic properties. Following Levelt and colleagues, this amodal grammatico-semantic level of representation is called the lemma level (Levelt 1993). In this case repeated activation of the stem should lead to faster responses on subsequent presentations for regular inflections only. Delayed priming studies in both the auditory (Kempley and Morton 1982) and visual (Napps 1989; Stanners et al. 1979) modalities found that irregular past tense forms primed their stems less than regular forms Sonnenstuhl et al. (1999) found a similar dissociation between regulars and irregular forms in German using cross-modal repetition priming.

However, studies in English, French, and Italian have found equivalent priming effects for regular and irregular inflections using masked (Forster et al. 1987; Meunier and Marslen-Wilson 2004), and cross-modal (Meunier and Marslen-Wilson 2004; Orsolini and Marslen-Wilson 1997) priming paradigms, suggesting that both regular and irregular inflection are represented in a single system.[1]

Other studies provide evidence that at least some irregular inflections prime their stem targets, but the studies come to conflicting conclusions about exactly which irregulars these are. Using the masked priming paradigm, Pastizzo and Feldman (2002) found priming for high overlap irregulars like gave-give, but not for low overlap pairs like taught-teach, while Kielar et al. (2008) and Allen and Badecker (2002a) found essentially the opposite pattern.

In an effort to resolve these conflicts, Crepaldi et al. (2010) compared morphologically related irregular prime-target pairs (e.g., *sold-sell*, *mice-mouse*) to pairs like *bold-bell* and *spice-spouse*, that share the same degree of orthographic overlap as the genuinely morphologically related pairs, but no semantic overlap. Any priming for the morphological pairs over and above that seen for the purely orthographically related pairs can be attributed to the repeated activation of a common morphological representation. Using the masked priming design, Crepaldi et al. found that irregularly inflected primes significantly facilitated reaction times to their targets, but that there was no facilitation for prime-target pairs where the prime and target were similar to one another in the same way as an existing irregular prime and target, but had no plausible morphological relationship to one another (e.g.,

[1] Note that French and Italian both offer a much less clear-cut distinction between regularity and irregularity than English in that irregular verbs typically also participate in many fully regular inflectional paradigms.

bold-bell). Crepaldi et al. did not directly compare regulars and irregulars. Their results, and the results of the other studies reporting significant masked priming effects for irregular allomorph primes, suggest that irregular allomorphs are very rapidly analyzed, leading to rapid activation of their stem correlates, and thus facilitation in the masked priming paradigm.

The lack of a pseudo-irregular priming effect in the Crepaldi et al. experiments raises doubts about whether irregulars are analyzed by the same morpho-orthographic mechanisms as regular allomorphs with overt, segmentable affixes, and poses difficulty for the models that propose an early morphological decomposition procedure that is blind to semantic information, but provides support for models of complex word recognition that include a level of morpho-semantic representation located "at some intermediate level between the morpho-orthographic segmentation stage and the semantic system" (Crepaldi et al. 2010, p. 91).

If morphologically complex words are processed via two distinct procedures—a search through the mental lexicon for irregular forms that must be memorized by virtue of the fact that their forms are unpredictable from the form of their bases, and a rule-governed affixation procedure which concatenates stems and affixes to produce forms that a fully transparent and segmentable—it is conceivable these processes are subserved by distinct neurophysiological mechanisms. To examine this possibility a number of studies have looked at ERP responses to regular and irregular verbs using priming tasks.

Münte et al. (1999), using a long-lag repetition priming paradigm, recorded ERPs as subjects read stem forms of regular and irregular verbs preceded by their past tense forms. They found that the ERPs to regular verbs primed by their past tense forms were associated with an N400 reduction while primed irregular verb stems showed no such effect. Similar dissociations in N400 priming effects between regular and irregular verbs were found in both German (Weyerts et al. 1996) and Spanish (Rodriguez-Fornells et al. 2002).

Kielar and Joanisse (2010), using a visual lexical decision task with both auditory and visual primes, compared N400 priming effects for regular verbs, irregular verbs with a stem change in the past tense form (e.g., sang-sing) and irregular verbs, such as *kept-keep*, which although irregular in that they involve an unpredictable stem change, also display a partial regularity in the form of a somewhat regular suffix (−*t*). They observed stronger N400 priming effects for regular than for vowel-change irregular verbs; the effects for the suffixed irregulars were similar to those for the regular verbs. They interpret their failure to observe a clear-cut regular/irregular dissociation as support for connectionist models that posit a graded rather than categorical distinction between regular and irregular forms.

All of these studies have examined the N400 component, which is sensitive to the mapping of lexical form onto meaning. However, if regular inflectional morphology is detected on the basis of morpho-orthograpic structure, priming effects are more likely to be detected in the N250, which reflects the process that maps sub-lexical orthographic and phonological representations onto whole word representations. In a recent study in our lab (Morris and Stockall 2012), using a masked priming paradigm, we examined response times as well as N250 and N400 responses to regular and irregular verbs each of which appeared in four conditions: primed by itself, primed by its past tense form, primed by an orthographic control that differed in only one letter, and primed by an unrelated item with which it shared no letters. We found robust facilitation effects for regular past tense priming in the N250 window. The response time data were similar to those of previous overt priming studies that have found full priming for regular verbs and reduced priming for irregulars. In both the N250 and the N400 time windows we found significant effects of the priming manipulation for both regular and irregular past tense primes, and no differences in the magnitude of the responses to the two types of target (but significant differences between morphological priming and orthographic overlap).

These data pose problems for both the Words and Rules and the connectionist models of morphological processing. The WR model would not predict an equivalent degree of lexical priming for both regular and irregular verbs given that the latter do not share a lexical entry with their stems, and connectionist models would only do so in cases where the degree of formal and semantic overlap between past tenses and stems is similar.

3.4.2 Frequency Effects

It has long been known that high-frequency words are recognized faster than low-frequency words (Forster and Chambers 1973), a phenomenon known as the word frequency effect. A line of research that attempts to address the question of whether morphologically complex words are represented and processed in decomposed form has focused on exploiting frequency effects in word recognition. Three separate frequency counts can be considered relevant in predicting recognition times for morphologically complex words, (1) the frequency of the specific surface form of the complex word, i.e., the *surface frequency*, (2) the frequency of the root and all affixed forms that share the same root, i.e., the *cumulative root frequency*, and (3) the number of distinct words containing the root as a constituent, i.e., the *type frequency of the root*, or *morphological family size.*

In studies using the visual lexical decision task in English (Taft 1979b) and Italian (Burani and Caramazza 1987) as well as a study using auditory lexical decision in French (Meunier and Segui 1999), it has been shown that both the surface frequency of a complex word itself, and the cumulative frequency of its stem, influence response times. However, the cumulative root frequency effect appears to be limited to suffixed words (Colé et al. 1989), consistent with a proposal by Taft that words are recognized via a left-to-right parsing process in which a search is undertaken for successive letter combinations beginning with the initial letter of the word (Taft 1979a, 1988). More recently, it has been shown that a larger morphological family size is associated with shorter latencies once frequency is controlled (Bertram et al. 2000; Schreuder and Baayen 1997). Schreuder and Baayen (1997) did not observe an effect of morphological family size using a progressive demasking task, a task that slows down the rate at which information becomes available and thus is primarily sensitive to lexical processing at the early stages of visual identification (Grainger and Segui 1990). Bertram et al. (2000) found that it is possible for complex words to show an effect of morphological family size without an effect of base frequency (the summed frequencies of the base word itself and its inflectional variants) and that the effect is driven only by the semantically transparent members of a morphological family.

The sensitivity of lexical decision times to both the surface and cumulative root frequencies of complex words is consistent with dual-route models that propose that the recognition of complex words can involve the activation of both whole-word representations and root representations. The absence of an effect of morphological family size in the progressive demasking task and the sensitivity of the morphological family size effect to semantic transparency suggest that that this effect arises due to activation spreading between semantically transparent morphologically related words stored in the central lexicon. Taken together these data are consistent with models in which not only is there decomposition at the level of form, but in which morphological relatives are linked to common morphological representations at a higher level of linguistic structure, with the patterns of connectivity determined by semantic transparency.

3.4.3 Semantic Transparency Effects

Morphological relatives tend to share both meaning and form. However, the meaning of a complex form is not always predictable from the meaning of its constituent morphemes and the degree to which morphologically complex words are semantically transparent may be important in determining whether they undergo decomposition. A morphologically complex word is semantically transparent if its meaning can be derived directly from the combined meaning of its stem and affix (e.g., goodness). In contrast, the meaning of a semantically opaque word (e.g., *department*) cannot be directly derived from the combined meaning of its component morphemes. Although at an earlier time in the history of the English language such words may have been related, and they may still be considered to have a morphological structure in that they consist of a clear stem and affix, the morphological relationship is orthographic rather than semantic.[2]

Using a cross-modal priming paradigm Marslen-Wilson et al. (1994) showed that there is significant priming for morphologically related words pairs that are semantically transparent (e.g., government-govern), but not for those that are semantically opaque (e.g., *apartment-apart*). These data suggest that semantically transparent items share the lexical entry of their common stem, and that lexical access is achieved via a representation of the stem rather than the derived form.

[2] In the literature, pseudocomplex words such as 'corner' that are in fact monomorphemic although their orthographic structure is such that they appear be composed of a stem and affix, tend to be included in the class of semantically opaque words.

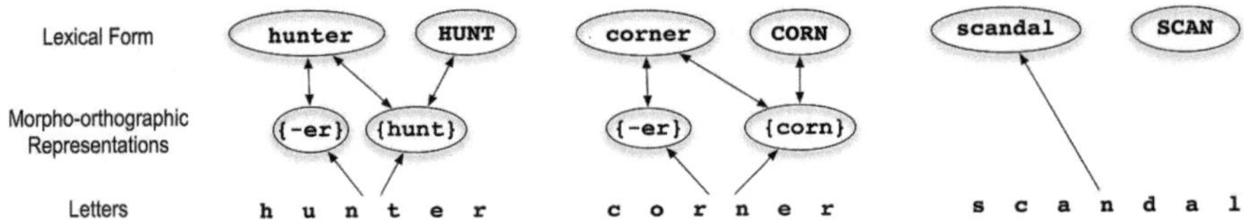

Fig. 3.4 A model of morpho-orthographic decomposition in which the recognition of morphologically complex words begins with a rapid morphemic segmentation based purely on the analysis of orthography

Subsequent studies confirmed this finding, but suggested that the advantage for transparent primes is apparent only with long (>200 ms) SOAs (Feldman and Soltano 1999; Rastle et al. 2000).

Studies using the masked priming paradigm confirm that the priming advantage for transparent morphological relatives is modulated by prime visibility.[3] With fully visible primes, primes and targets must show semantic as well as formal overlap in order to induce a priming effect. In contrast, in the masked priming paradigm semantically opaque primes show priming effects over and above those shown by matched orthographic control primes (Gold and Rastle 2007; Lavric et al. 2007; Longtin et al. 2003; McCormick et al. 2008; Rastle et al. 2004), but only at very short prime durations (Dominguez et al. 2002; Rastle et al. 2000). This is clear evidence for a form of automatic morpho-orthographic decomposition operating during the early phases of visual word recognition that is sensitive to the superficial morphological structure of strings, but blind to the semantic consequences of the decomposition (see Fig. 3.4).

Because morpho-orthographic decomposition has been argued to be a sub-lexical phenomenon, it should be unaffected by the lexical status of the prime. In a masked priming study conducted in French, Longtin and Meunier (2005) confirmed that indeed, morphologically complex pseudowords significantly facilitated the recognition of their embedded stems (e.g., *rapidifier-rapide*), as much as did existing derived words (e.g., *rapidement-rapide*). However, no priming effect was obtained with non-morphological pseudowords (e.g., *rapiduit-rapide*), in which the orthographic string following the stem (e.g., "*–uit*" in the preceding example) is not a valid suffix in the language. These data support the contention that decomposition is based on morpho-orthographic rather than strictly orthographic units.[4] Moreover,

[3] Morphological priming effects do not appear to vary with semantic transparency or prime visibility in Hebrew or Arabic, languages with a non-concatenative morphology (Boudelaa and Marslen-Wilson 2001; Frost et al. 2000b; 1997). The cause of this difference is unclear.

[4] In a recent study conducted in our lab, (Morris et al. 2011) we found statistically equivalent behavioral priming for derived word primes (flexible-flex), complex non-word primes (flexify-flex), and simplex non-word primes (flexint-flex). The key divergence with respect to Longtin and Meunier's results was our finding of priming with simplex nonwords, a condition that generated no priming in their experiments. One explanation for the discrepancy may be that in our stimulus set the target was always fully contained in the prime (e.g., flexify-FLEX). However, in the stimulus set used by Longtin et al., the orthographic overlap between primes and targets was not always complete (e.g., chambrage-CHAMBRE). It may be that for simplex primes, complete overlap is required to activate the embedded target string.

the syntactic legality and semantic interpretability of these morphologically structured pseudoword primes had no influence on masked priming effects.

There are data to suggest that the representation of morpho-orthographic units, although formal, may be quite abstract. McCormick et al. (2008) showed that morphological priming occurs even in the face of common orthographic alterations found at morpheme boundaries in complex words. These include a missing "e" (e.g., *adorable–adore*), a shared "e" (e.g., *lover–love*), and a duplicated consonant (e.g., *dropper–drop*). Moreover, it was shown that this robustness can be found even when the relationship between the prime and the target is semantically opaque (e.g., fetish-fete). These data suggest that pre-lexical morph-orthographic units may be represented in an abstract underspecified form.

If semantically transparent and semantically opaque complex words both undergo a common morpho-orthographic decomposition, it is likely that this process is subserved by a common neurophysiological mechanism. To examine this possibility a number of studies have looked at ERP responses to word stems primed either by semantically transparent morphological relatives (*hunter-hunt*), pseudo-complex (and hence semantically opaque) relatives (*corner-corn*) or words that share a purely orthographic relationship (*scandal-scan*). If both transparent and opaque words undergo a similar decomposition process we should find ERP responses that are similar for these two conditions, but differ from those for the purely orthographic condition.

Lavric et al. (2007) examined both response times and the mean amplitude of the N400 component to targets preceded by words with a transparent morphological, pseudo-morphological or purely orthographic relationship. The behavioral data were similar to those found in previous studies of semantic transparency using the masked priming paradigm (Longtin et al. 2003; Rastle et al. 2004), i.e., significant priming effects of roughly equal magnitude for the transparent and opaque conditions, and significantly less priming for the orthographic condition. The ERP data revealed a similar pattern, in that a reduction in the amplitude of the N400 component appeared later and was more short-lived for the orthographic condition than for the transparent and opaque conditions. In an MEG study of masked morphological priming similar to that of Lavric et al. (2007), Lehtonen et al. (2011) measured the latency of an MEG component peaking, on average, 220 ms post-onset of the target in left occipito-temporal brain regions. Like Lavric et al. (2007), they found significant priming effects (shorter latencies for related relative to unrelated pairs) for transparent and opaque prime-target pairs, but not for pairs with a purely orthographic relationship.

These data are consistent with prior priming studies using long stimulus onset asynchronies (SOAs) in Spanish (Barber et al. 2002; Dominguez et al. 2004) which compared ERP responses to stem homographs and genuine morphological prime-target pairs. In Spanish, gender is marked by adding the suffix "-a" to root a morpheme obtain a feminine word, or the suffix "-o" to obtain a masculine word, (e.g., *loc-a* (madwoman)/*loc-o* (madman)). Stem homographs are words with stems that are orthographically identical, but semantically and morphologically unrelated. For example, in Spanish, *rat-a* ("rat," feminine singular) and *rat-o* ("moment," masculine singular) are related neither semantically nor morphologically, yet their stems (*rat-*) are orthographically identical. Unlike morphologically related inflections that produce facilitation in priming tasks, stem homographs produce inhibition, at least at long SOAs (Dominguez et al. 2002; Laudanna et al. 1989; Laudanna et al. 1992). Barber et al. (2002) found attenuation of the N400 component for both homographic and morphological prime-target pairs. In a more carefully controlled study, Dominguez et al. (2004) found that morphologically related pairs elicited a broad attenuation of the N400. In contrast, stem homographs were associated with a reduction in the amplitude of the N400 in the early phase of the N400 time window (250–350 ms), followed by an increase in the N400 amplitude in a later phase (350–450 ms).

Lavric et al. (2011) conducted a similar study examining the ERP correlates of semantic transparency in a morphological priming study with briefly presented but fully visible primes in English. They found that priming effects—characterized by a reduction in the N400 component—for semantically transparent (*hunter-hunt*) and semantically opaque (*corner-corn*) morphologically structured prime-target pairs were equal to each other and both greater than the effect for purely orthographic priming in the early part of the N400 time range (300–380 ms), reminiscent of the N400 results. Lavric et al. (2007) obtained previously with masked priming. In the latter phase of the N400 (from 380 ms onward), the priming effect (N400 attenuation) for opaque pairs was reduced to the level seen in the orthographic condition, whereas N400 priming in the transparent condition continued to be robust.

The data from these three studies (Barber et al. 2002; Dominguez et al. 2004; Lavric et al. 2011) are consistent with a processing model in which an initial morpho-orthographic decomposition results in activation of an orthographic representation of stem that is the same for stem homographic word pairs. When semantic information becomes activated at a later stage, the incompatibility between the meaning of the embedded root and the meaning of the complete prime word, results in the reduced priming effects found for these items in the late N400 component. Taken together, these data provide support for a model of complex word recognition in which a morphologically structured level of representation plays an important role and suggest that early morphological influences can be obtained independently of semantic relatedness which should only influence morphological priming at longer prime durations.

However, the evidence in favor of a pure morpho-orthographic decomposition process is not unequivocal. Although reports of significant effects of semantic transparency are few compared to the number of published studies that have reported non-significant differences between semantically transparent and semantically opaque morphological primes, a recent meta-analysis of 16 published experiments that manipulated semantic transparency with masked morphological primes (Feldman et al. 2009) show that the effect of semantically transparent primes is significantly larger than that of opaque primes. And indeed, some individual studies using masked priming studies have shown a statistically significant, as opposed to merely numerical, advantage for semantically transparent derived primes relative to semantically opaque and/or pseudo-derived primes (Diependaele et al. 2005, 2009; Feldman et al. 2009; Morris et al. 2007).

There is also neurophysiological evidence for semantic transparency playing a role in morphological processing. Kielar and Joanisse (2011) measured morphological priming effects in the N400 ERP component during a cross-modal lexical decision task for prime-target pairs with unambiguously transparent (government–govern), unambiguously opaque (apartment–apart), and intermediate (dresser–dress) semantic relationships. The degree of N400 ERP priming was modulated by the amount of semantic and phonological overlap between prime and target.

Evidence for the early involvement of semantics comes from a lexical decision masked priming study conducted in our lab (Morris et al. 2007). In this study we compared mean N250 and N400 amplitudes, as well as behavioral responses to targets primed by semantically transparent, semantically opaque, and orthographically, but not morphologically, related primes. We found items that in the transparent condition generated greater priming effects in all three dependent measures than did items in the orthographic and opaque conditions. Furthermore, comparison of the mean differences between unrelated and related items across conditions revealed the existence of a significant linear trend, with transparent items showing the greatest effects and orthographic items the smallest, suggesting that these priming effects vary as a function of morphological structure and semantic transparency. In a subsequent study using a semantic categorization task (Morris et al. 2008) we found a widespread priming effect in the early phase of the N250 for both morphologically related semantically transparent and semantically opaque morphological primes, but a dissociation between semantic and opaque primes in the late phase of the N250.

These data support the hypothesis that while early in visual word recognition there is a process of morpho-orthographic segmentation that operates independently of the semantic relatedness of the embedded root and the whole-word form, semantic transparency may modulate lateral inhibitory connections between whole-word form representations, i.e., there may be reduced competition between words that have both a morphological and a semantically transparent relation, and this may be reflected in the later phase of the N250.

Sub-lexical morpho-orthographic segmentation may be only one of the pathways within a dual-route model of morphological processing that includes a role for both orthographic and semantic representations in the early stages of word recognition (Diependaele et al. 2012; Diependaele et al. 2013; Diependaele et al. 2005, Diependaele et al. 2009; Morris et al. 2011). This proposal suggests that for complex words, representations at the lexical level can be activated not only via sub-lexical morpho-orthographic representations, but also by a route that maps the whole input directly onto a whole-word form representation. Lexical representations in turn map onto higher-level morpho-semantic representations.

Evidence for this view comes from a study by Diependaele et al. (2012) that examined transposed letter effects in complex words. This study draws upon the work of Grainger and colleagues who have recently proposed a model of word recognition that draws a distinction between two types of sub-lexical orthographic codes, a coarse-grained code that provides fast access to semantic representations by using subsets of letters that allow for rapid word identification, and a fine-grained code in which attention is paid to the precise ordering of letters and is used for detecting the highly frequent letter combinations that are common in multi-letter graphemes and affixes (Grainger and Ziegler 2011). The coarse-grained code maps letter representations onto whole word representation and then onto morpho-semantic representations. The fine-grained code, in contrast, maps letter representations onto sub-lexical morpho-orthographic representations. Thus letter transpositions, that only mildly impair fluent reading (Rayner et al. 2006), should disrupt morpho-orthographic processing more than morpho-semantic processing. Since morphological priming effects for semantically opaque or pseudo-morphological relatives depend exclusively on the sub-lexical morpho-orthographic route, letter transpositions should eliminate this priming effect. In contrast, because semantically transparent morphological relatives share supra-lexical morpho-orthographic representations that can be accessed via the coarse-grained code, priming effects for semantically transparent morphological relatives should be preserved. This is indeed what was found—no priming from opaque pseudo-derived primes containing a letter transposition at the morpheme boundary (*masetr-mast*) in the presence of significant priming from transposed-letter transparent derived primes (*banekr-bank*).

The preponderance of the evidence from behavioral, electrophysiologial, and neuroimaging studies examining morphological priming and root frequency effects suggests that lexical representations contain explicit representations of morphological structure, and that morphological parsing may be early and pre-lexical, based on orthographic rather than semantic information. In the following section, I will examine the neurophysiological evidence for morphological decomposition and consider whether the processes responsible for both morphological and syntactic parsing share a common brain mechanism.

3.5 Neurophysiological Markers of Morphological Parsing

Given that both morphology and syntax are systems that govern the expression of linguistic form, and that both systems are combinatorial and rule governed, it may be that the cognitive processes responsible for the decomposition and/or combination of both morphological and syntactic units share a single underlying brain mechanism. Indeed, Marantz (2013) makes a theoretical argument that "within the theoretical framework of Distributed Morphology … the internal structure of words finds analysis within the same syntactic architecture and subject to the same syntactic principles as the internal structure of phrases and sentences" (Marantz 2013, p. 906).

Electrophysiological techniques have been shown to be uniquely well suited to elucidating the relationship between cognitive processes and brain function. In particular, there are a number of ERP components that have been shown to be sensitive to expectancy violations in the syntactic domain. The P600 is a large positive waveform with a mainly posterior scalp distribution, peaking at approximately 600 ms after the onset of words that are inconsistent with the expected structural analysis of a sentence. A disparate set of syntactic violations have been shown to enhance the amplitude of the P600. These include violations of constraints on phrase structure (Hagoort et al. 1993; Neville et al. 1991; Osterhout and Holcomb 1992), verb subcategorization (Hagoort et al. 1993; Osterhout and Holcomb 1992; Osterhout et al. 1994; Rösler et al. 1993), subject–verb agreement (Hagoort et al. 1993; Osterhout and Mobley 1995; De Vincenzi et al. 2003), reflexive pronoun–antecedent agreement and anaphor binding (Harris et al. 2000; Osterhout and Mobley 1995) and subjacency (McKinnon 1996; Neville et al. 1991), as well as grammatical but non-preferred continuations (Kaan et al. 2000; Kaan and Swaab 2003). In addition to the P600, violations of syntactic rules such as errors of subcategorization and to morphological agreement violations, e.g., subject–verb agreement lead to a left anterior negativity (LAN) in the same time window as the N400 (Friederici 1995; Rösler et al. 1993).

Thus, if the cognitive processes responsible for the decomposition and/or combination of both morphological and syntactic units share a single underlying brain mechanism, modulation of components such as the LAN and P600 may provide evidence for a rule-based combinatorial accounts of morphological representation and processing. Several studies have reported LAN effects in response to violations of morphological rules. Penke et al. (1997) examined ERPs to violations of regular and irregular verb inflection in German, created by means of a "regularity" mismatch between stem and affix; *–(e)n* was attached to verbs that take *-t* participles *(*getanz-en)*, and *-(e)t* to verbs that take *-(e)n* participles *(*gelad-et)*. They found a LAN for irregular verbs with the incorrect regular *-t* participle (e.g., **aufgeladet* versus *augeladen*) but not for regular verbs, with the irregular *-n* participle (e.g., **durchgetanzen* versus *durchgetanzt*). Weyerts et al. (1996), reported similar findings for regular and irregular noun plurals. Rodriguez-Fornells and colleagues found both a LAN, albeit one with a more posterior distribution than has typically been found, and a P600 to verb stem formation errors, and past tense inflectional over-regularization errors in Catalan (Rodriguez-Fornells et al. 2001).

However, other studies have failed to find LAN effects for morphological violations presented as single words and not embedded in sentential contexts. Morris and Holcomb (2005) found that morphological violations consisting of incorrect stem/affix combinations presented in a sentence context elicited a bi-phasic response consisting of a LAN followed by a P600, while similar violations presented as single words led to a N400 followed by a P600.

These findings suggest that the presence of a P600 effect in response to morphological violations may reflect difficulties with combinatorial processes at multiple levels of linguistic analysis, i.e., both morphological and syntactic. In particular, the P600 may reflect the detection of a mismatch between syntactic or morphological units, or it may signal the reanalysis of such a mismatch. In contrast to the P600, the LAN only appeared as a response to violations in sentence contexts. Thus, the LAN may be sensitive to processes involved in integrating a word into its syntactic context, as opposed to the P600 which may reflect morphosyntactic processes that can be, although they are not necessarily, independent of a sentential context (e.g., reanalysis of the morpho-syntactic structure of a word or phrase).

3.5.1 Direct Measures of Morphological Parsing

In morphological priming studies, morphological processes operating on complex prime stimuli are inferred as a result of the subsequent effects of such processes—either facilitation or inhibition—on the recognition of related simplex targets. In violation studies, differences in the timing or strength of responses between legal and illegal word or morpheme combinations are presumed to reflect the disruption of morphological and syntactic rules. Yet, despite the interest in the neural substrates of morphological and syntactic parsing there are relatively few studies that have attempted to directly identify or measure neural activity related to morphological parsing.

One study by Lehtonen et al. (2007) contrasted inflected and monomorphemic Finnish words and found an effect of complexity only in the N400 time-window, with a greater negativity for inflected words. Yet, the timing of this effect is surprising. If decomposition is pre-lexical and precedes semantic access, we should observe a much earlier effect of morphological structure.

Early effects of morphological complexity have been reported in studies using MEG. Zweig and Pylkkänen (2009) examined MEG responses to morphologically structured words consisting of real stems plus real affixes that were either semantically transparent (*farmer*) or semantically opaque (*sweater*). They also examined responses to words comprising a non-word stem plus real affix (*winter*), and simplex words with no stems or affixes *switch*). They found effects of complexity on the M170, a MEG component that appears to originate in inferior temporal cortex including the visual word form area (VWFA) at about 150 ms post-stimulus (Solomyak and Marantz 2009). Morphologically structured words elicited larger M170 amplitudes than did simplex words or words comprising a non-word stem plus real affix. The sensitivity of the M170 to morphological complexity has been confirmed in subsequent studies by Solomyak and Marantz (2010) and Lewis et al. (2011) who showed that that the M170 is sensitive to the morphological, rather than to purely orthographic properties of complex words. Moreover, for pseudo-affixed words like *brother*, the statistical relationship between stem and affix modulates the M170 response as much as for truly affixed words.

Early effects of morphological structure have also been recently reported in the EEG literature. Lavric et al. (2012) directly compared processing of complex, pseudocomplex, and simplex words in an unprimed lexical decision task with ERP recordings. They found that the ERP waveforms patterned together for complex and pseudocomplex words in an early time-window (190–220 ms post-stimulus onset), both differing from the simplex word ERPs. In a slightly later time window (246–

276 ms post-stimulus onset) ERPs for the complex and simplex words patterned together, and differed from those for the pseudocomplex words. These studies yield evidence for a form of early morphological decomposition based initially on orthographic analysis that operates on complex and pseudocomplex words but not simplex words, and is only later constrained by semantic information.

A recent study from our lab (Morris et al. 2013) examined responses to complex, pseudocomplex target words, and simplex target words preceded by complex nonword primes, in order to directly examine ERP responses to morphological decomposition as it occurs, during the processing of target words of varying levels of complexity. The complex targets were primed by both related and unrelated complex nonword primes formed of a combination of the stem of the target and a suffix that did not generate an existing word (e.g., farmity-farmer). The rationale is that these nonword primes will induce morpho-orthographic decomposition (Longtin and Meunier 2005), and we should be able to trace the influence of this morpho-orthographic decomposition during the processing of the complex target in the absence of any morpho-semantic priming.

For the unrelated prime conditions (equivalent to an unprimed study) in the early 150–200 ms time window, we found that ERPs to complex and pseudocomplex targets patterned together, and both were significantly less negative-going than the ERPs to simplex targets. This result is in line with the pattern reported by Lavric et al. (2012), as well as the results from MEG studies (Lewis et al. 2011; Solomyak and Marantz 2010) showing that the M170 component increases as a function of the decomposability of words. In the N400 epoch, on the other hand, we found that pseudocomplex words patterned with simplex words and these both generated more negative-going waveforms than complex words.

In the related priming conditions, we found a significant priming effect that was only seen for semantically transparent complex target words in the earliest time-window, between 150 and 200 ms post-target onset. Following that, all types of target showed widely distributed priming effects in the time window typically associated with the N250 ERP component, and following that, more spatially focused priming effects in the N400 time window. The existence of this early priming effect limited to transparent words suggests that although the process of morpho-orthographic segmentation might be blind to whether or not the stimulus is truly morphologically complex, the subsequent use of the segments extracted by this process is determined by whether or not the stimulus is truly complex. That is, although "corn" may be extracted "cornity" by a process of morpho-orthographic decomposition, that representation of "corn" will not prime "corner." In contrast, the representation of "hunt" extracted from the non-word "huntity" will prime the truly complex word "hunter."

Taken as a whole, the data show that masked priming effects vary as a function of both morpho-orthographic structure and semantic transparency. These results are difficult to reconcile with either a prelexical decomposition account or a purely supralexical account of morphological representation, but rather suggest that there are two distinct sources of morphological influences on visual word recognition—morpho-orthographic decomposition and morpho-semantic processing.

Morpho-orthographic decomposition is a very early sub-lexical process that relies on a fine-grained code and is blind to semantic information; this is the process that yields morphological priming effects for both semantically transparent and semantically opaque morphologically structured words. At the same time, whole-word form representations are activated via a coarse-grained orthographic code, and in turn activate supra-lexical morphological representations that code for semantically transparent morphological relations between words.

These processes are reflected in the neurophysiological data. In particular, the N250 ERP component is affected by semantic transparency, reflecting a state of resonance between prelexical (bottom-up) and lexical–semantic (top-down) representations, and not just bottom-up prelexical activation or only top-down semantic activation. When a semantically transparent prime (e.g., *hunter*) activates its embedded stem (*hunt*), the subsequent processing of the target "hunt" benefits from the compatibility between activated prelexical form representations and higher level semantic representations. This does not occur with opaque and orthographic primes, given the semantic incompatibility between the prime word's meaning and the meaning of the target word. Semantic transparency also influences later processing reflected by the N400 ERP component and overt behavioral responses.

3.6 Neuroanatomical Substrates of Morphological Processing

As noted previously, the combinatorial versus the associative or lexical nature of morphological processing has been hypothesized to hinge on various properties of morphological representations such as whether affixes are inflectional or derivational, regular or irregular, productive or unproductive or semantically transparent or opaque. As a result, much of the research on the neuroanatomical substrates of morphological processing has focused on finding areas involved in morphological processing, using functional magnetic resonance imaging (fMRI) tasks that reveal distinctions between these classes of complex words.

3.6.1 Regularity

Ullman has explicitly linked the "words and rules" model to the contrast between procedural and declarative memory systems (Ullman 2001, 2004; Ullman et al. 1997). According to the declarative-procedural (DP) model, lexical memory depends upon declarative memory, subserved by temporal lobe structures, while combinatorial rule-based language systems such as syntax, regular inflectional and derivational morphology and aspects of phonology, depends upon procedural memory, sub-served by left frontal/basal-ganglia structures. If irregular forms are memorized, while regular forms are generated by a rule, any impairment of lexical memory from damage to temporal or parietal neocortex should affect irregular forms more than regular. In contrast, impairments of procedural memory from damage to frontal cortex or the basal ganglia should lead to difficulty in producing regular forms. Ullman et al. (1997) showed that patients with temporal or parietal damage and impairments of declarative memory as a result of Alzheimer's disease had more difficulty producing irregular than regular past tense verb forms. In contrast, Parkinson's disease patients with predominantly frontal lobe or basal ganglia damage showed the opposite pattern, i.e., difficulty producing regular past tenses.

Functional imaging studies using non-patient populations have also reported distinct patterns of activation for regular and irregular past tense verb forms. Jaeger et al. (1996) conducted a positron emission tomography (PET) study in which subjects were asked to produce the past tense forms of regular and irregular verbs. They found differences in both the location and the amount of brain activation in the regular vs. irregular tasks, with a much larger area of activation for irregulars. More specifically, for regular inflection they found activation of the left dorsolateral prefrontal cortex and the left anterior cingulate cortex. They argue that as this area is involved in intentional and rule-governed, as opposed to automatic stimulus-driven behavior, it may be an area involved in regular suffixation. In contrast, for irregular past tenses, they found activation of the left middle temporal gyrus, an area associated with long-term semantic memory. Activation of this area lends support to the hypothesis that irregular forms are retrieved from memory rather than computed online. Furthermore, for irregular inflections they also found activation of the left-lateral orbito-frontal cortex, an area involved in inhibiting automatic learned responses to stimuli; this is consistent with the claim that regular affixation must be blocked in order for irregular past tense forms to be retrieved.

Although both Ullman et al. (1997) and Jaeger et al. (1996) found that distinct neuroanatomical networks subserve regular and irregular inflection, the specific areas that were activated for the two verb types differed in the two studies. Jaeger et al., unlike Ullman et al., did not find significant activation of the basal ganglia for regular past tenses. Moreover, although Ullman et al. argue that Broca's area is part of the network for regular inflection, Jaeger et al. found that this area was activated for production of both regular and irregular past tenses.

In an fMRI study of German regular and irregular inflection, Beretta et al. (2003) provided further support for differing patterns of activation for regular and irregular forms. In this study, as in the Jaeger et al. (1996) study, irregular words showed a wider extent of activation than did regular words, suggesting that searching the lexicon for an irregular past tense form costs more in terms of neural resources than mere suffixation. However, while Jaeger et al. found greater left hemisphere activation for both regulars and irregulars, Beretta et al. found a hemispheric split in activation between the two verb types; areas activated by regulars showed more left hemisphere activation than right, whereas areas activated during irregular inflection showed either no significant difference between hemispheres, or a tendency toward right lateralization. They explain this difference in activation patterns by proposing that irregular forms must be retrieved from memory and such retrieval may be aided either by information stored in specific lexical entries in semantic memory, or by information stored in specific episodic memories of past language experiences. If retrieval from semantic memory tends to activate left prefrontal cortex more than right, while retrieval from episodic memory activates right prefrontal cortex more than left, when both forms of memory are involved bilateral activation results.

These studies are consistent with the claim that there are distinct processing systems subserving regular and irregular inflectional morphology, that regular inflection activates areas implicated in rule-governed behavior and that irregular inflection is in some sense more effortful involves response inhibition and shows neural signatures of retrieval from memory.

3.6.2 Inflection Versus Derivation

In an early fMRI study, Miceli et al. (2002) examined the neuroanatomical substrates of morphological, semantic, and phonological processing by presenting participants with a target word and asking them to make judgments about its morphological (grammatical gender), semantic, or phonological features. They found that morphological judgments activated areas of the left inferior frontal and temporal cortex, highlighting the critical role of the left frontal lobe in the processing of inflectional morphology. Subsequent studies by Tyler et al. (Tyler, Marslen-Wilson, & Stamatakis, 2005; Tyler, Stamatakis, Post, Randall, & Marslen-Wilson, 2005) also found that priming effects for regularly inflected verbs correlated with activation of left frontotemporal areas. Thus, left frontotemporal regions appear to be implicated in the processing of inflectional morphemes.

In a more recent fMRI study, Bozic et al. (2013) used a gap detection paradigm to examine derived words that differed in terms of their semantic transparency and affix productivity. Their goal was to compare the pattern of activation for the kinds of derived words most likely to be processed via decomposition, i.e., semantically transparent words with productive affixes—to that of regularly inflected words which have been shown to activate a left hemisphere frontotemporal system hypothesized to support grammatical computations (Bozic et al. 2010). If derived words are stored in decomposed form and processed via to their constituent morphemes, they should activate the left hemisphere frontotemporal system, in a manner similar to that of inflected words. They found no activation of the LH frontotemporal system for derived words, but rather observed activation in bilateral middle temporal gyrus (MTG) and in right inferior frontal gyrus (RIFG), a network that has been implicated in lexical processing and the mapping between sound and meaning. These findings suggest that derived and inflected words are sub-served by distinct neuroanatomical systems with only the latter engaging a left hemisphere frontotemporal system specialized for processing grammatical language functions.

3.6.3 Productivity and Semantic Transparency

Studies examining the neuroanatomical substrates of morphological decomposition in derivation have focused on contrasting both productive and unproductive and semantically transparent and opaque forms on the assumption that productive, transparent forms are more likely to be decomposed that unproductive opaque ones. In an fMRI study, Vannest et al. (2005) compared words with productive suffixes (e.g., *−ness*, *−less*, and *–able)* that show evidence of decompositional processing in behavioral studies (Vannest et al. 2002; Vannest and Boland 1999) to words with non-productive suffixes (e.g., *−ity* and *–ation*), that have not shown decomposition effects. They found that words with productive suffixes showed a greater pattern of activation in Broca's area and the basal ganglia than words with non-productive affixes, similar to the pattern predicted by Ullman et al. (1997) for regular inflection. These data suggest that the "productive/unproductive" distinction in derivation may parallel that of the "regular/irregular" distinction in inflection, and the processing mechanisms hypothesized to underlie this distinction may extend to morphologically complex forms created by derivational processes.

In order to test the hypothesis that complex word recognition involves a rapid morphemic segmentation based purely on the analysis of orthography and blind to semantic representations, Gold and Rastle (2007) used a masked priming visual lexical decision task to look for neural markers of priming for semantically opaque derivations. They used as their dependent measure fMRI priming suppression, i.e., a reduced blood oxygen level dependent (BOLD) response to repeated stimuli. They found that morphological priming was observed in two regions that also showed orthographic priming, the posterior portion of the fusiform gyrus and a posterior portion of the middle occipital gyrus, but that the morphological priming regions did not overlap with those that showed activation in the semantic priming condition. Gold and Rastle interpret these data as supporting a morpho-orthographic segmentation mechanism that occurs independently of lexical–semantic processes, at a very early stage of visual word recognition.

3.7 Summary

The preponderance of the evidence from behavioral, electrophysiologial, and neuroimaging studies suggests that lexical representations are indeed morphologically structured, and that in reading, morphologically complex words are decomposed and analyzed in terms of their constituents; in other words, lexical representations contain explicit representations of morphological structure. Furthermore, the evidence suggests that morphological decomposition may be early and pre-lexical, based on orthographic rather than semantic information.

However, the consistent, if small, effects of semantic transparency on morphological priming for derivations (Feldman et al. 2009), and significant morphological priming effects for irregular inflections relative to orthographic controls (Crepaldi et al. 2010; Morris and Stockall 2012) suggests that the internal morphological structure of complex words may be represented at two locations in a hierarchically organized processing framework—a morpho-orthographic sub-lexical level that is used to parse the stimulus input, and a morpho-semantic, amodal supra-lexical level that codes for the similarity in form and meaning between semantically transparent morphologically related words.

Words and morphemes are the linguistic units that provide the main mapping between form and meaning, and yet few models of lexical processing have included morphology as a level of linguistic representation. Expanding the focus of research to include multi-morphemic words has significantly advanced our understanding of reading by allowing us to evaluate the contribution of morphological analysis to the process of word recognition. Moreover, greater understanding of the differences between combinatorial processes at the level of the word and at the level of the phrase, and between the principles that govern the composition of complex words and those that govern the composition of phrases will help to refine our understanding of the representations and processes that underlie our ability to both produce and comprehend complex words and phrases.

References

Allen, M., & Badecker, W. (2002a). Inflectional regularity: Probing the nature of lexical representation in a cross-modal priming task. *Journal of Memory and Language, 46*, 705–722.

Allen, M., & Badecker, W. (2002b). Stem homographs and lemma level representations. *Brain and Language, 81*, 79–88.

Aronoff, M. (1983). A decade of morphology and word formation. *Annual Review of Anthropology, 12*, 355–375.

Aronoff, M., & Fudeman, K. (2010). *What is morphology?: Fundamentals of linguistics* (1st ed.). Chichester: Wiley.

Baayen, R. H., Dijkstra, T., & Schreuder, R. (1997). Singulars and plurals in Dutch: Evidence for a parallel dual-route model. *Journal of Memory and Language, 37*, 94–117.

Barber, H., Dominguez, A., & de Vega, M. (2002). Human brain potentials indicate morphological decomposition in visual word recognition. *Neuroscience Letters, 318*, 149–152.

Beretta, A., Campbell, C., Carr, T. H., Huang, J., Schmitt, L. M., Christianson, K., & Cao, Y. (2003). An ERfMRI investigation of morphological inflection in German reveals that the brain makes a distinction between regular and irregular forms. *Brain and Language, 85*, 67–92.

Bentin, S., & Feldman, L. (1990). The contribution of morphological and semantic relatedness to repetition priming at short and long lags: Evidence from hebrew. *The Quarterly Journal of Experimental Psychology Section A, 42*, 693–711.

Bentin, S., & Frost, R. (1995). Morphological factors in visual word identification in Hebrew. In L. B. Feldman (Ed.), *Morphological aspects of language processing* (pp. 271–292). Hillsdale, NJ: Lawrence Erlbaum Associates, Inc.

Bertram, R., Baayen, R. H., & Schreuder, R. (2000). Effects of family size for complex words. *Journal of Memory and Language, 42*, 390–405.

Boudelaa, S., & Marslen-Wilson, W. (2001). Morphological units in the Arabic mental lexicon. *Cognition, 81*, 65–92.

Boudelaa, S., & Marslen-Wilson, W. D. (2005). Discontinuous morphology in time: Incremental masked priming in Arabic. *Language & Cognitive Processes, 20*, 207–260.

Bozic, M., Tyler, L. K., Ives, D. T., Randall, B., & Marslen-Wilson, W. D. (2010). Bihemispheric foundations for human speech comprehension. *Proceedings of the National Academy of Sciences of the United States of America, 107*, 17439–17444.

Bozic, M., Tyler, L. K., Su, L., Wingfield, C., & Marslen-Wilson, W. D. (2013). Neurobiological systems for lexical representation and analysis in English. *Journal of Cognitive Neuroscience, 25*, 1678–1691.

Burani, C., & Caramazza, A. (1987). Representation and processing of derived words. *Language & Cognitive Processes, 2*, 217–227.

Butterworth, B. (1983). Lexical representation. In B. Butterworth (Ed.), *Language production: Development, writing and other language processes* (Vol. 2). Cambridge, MA: Academic.

Caramazza, A., Laudanna, A., & Romani, C. (1988). Lexical access and inflectional morphology. *Cognition, 28*, 297–332.

Chomsky, N., & Halle, M. (1968). *The sound pattern of English*. New York: Harper & Row.

Clahsen, H. (2003). Derivational morphology in the German mental lexicon: A dual mechanism account. In H. Clahsen, I. Sonnenstuhl, & J. P. Blevins (Eds.), *Morphological structure in language processing* (Vol. 151, p. 125).

Colé, P., Beauvillain, C., & Segui, J. (1989). On the representation and processing of prefixed and suffixed derived words: A differential frequency effect. *Journal of Memory and Language, 28*, 1–13.

Crepaldi, D., Rastle, K., Coltheart, M., & Nickels, L. (2010). 'fell' primes 'fall', but does 'bell' prime 'ball'? Masked priming with irregularly-inflected primes. *Journal of Memory and Language, 63*, 83–99.

Diependaele, K., Grainger, J., & Sandra, D. (2012). Derivational morphology and skilled reading. In *The Cambridge handbook of psycholinguistics* (pp. 311–332). Cambridge: Cambridge University Press.

Diependaele, K., Morris, J., Serota, R. M., Bertrand, D., & Grainger, J. (2013). Breaking boundaries: Letter transpositions and morphological processing. *Language & Cognitive Processes, 28*, 988–1003.

Diependaele, K., Sandra, D., & Grainger, J. (2005). Masked cross-modal morphological priming: Unravelling morpho-orthographic and morpho-semantic influences in early word recognition. *Language & Cognitive Processes, 20*, 75–114.

Diependaele, K., Sandra, D., & Grainger, J. (2009). Semantic transparency and masked morphological priming: The case of prefixed words. *Memory & Cognition, 37*, 895–908. doi:10.3758/mc.37.6.895.

Dominguez, A., De Vega, M., & Barber, H. (2004). Event-related brain potentials elicited by morphological, homographic, orthographic, and semantic priming. *Journal of Cognitive Neuroscience, 16*, 598–608.

Dominguez, A., Segui, J., & Cuetos, F. (2002). The time-course of inflexional morphological priming. *Linguistics, 40*, 235–260.

Drews, E., & Zwitserlood, P. (1995). Morphological and orthographic similarity in visual word recognition. *Journal of Experimental Psychology: Human Perception and Performance, 21*, 1098.

Elman, J. L. (2004). An alternative view of the mental lexicon. *Trends in Cognitive Sciences, 8*, 301–306.

Feldman, L. B. (2000). Are morphological effects distinguishable from the effects of shared meaning and shared form? *Journal of Experimental Psychology Learning Memory and Cognition, 26*, 1431–1444.

Feldman, L. B., & Bentin, S. (1994). Morphological analysis of disrupted morphemes: Evidence from Hebrew. *The Quarterly Journal of Experimental Psychology, 47*, 407–435.

Feldman, L. B., & Moskovljević, J. (1987). Repetition priming is not purely episodic in origin. *Journal of Experimental Psychology: Learning, Memory, and Cognition, 13*, 573.

Feldman, L. B., O'Connor, P. A., & del Prado Martín, F. M. (2009). Early morphological processing is morphosemantic and not simply morpho-orthographic: A violation of form-then-meaning accounts of word recognition. *Psychonomic Bulletin & Review, 16*, 684–691.

Feldman, L. B., & Soltano, E. G. (1999). Morphological priming: The role of prime duration, semantic transparency, and affix position. *Brain and Language, 68*, 33–39.

Forster, K. I. (2009). The intervenor effect in masked priming: How does masked priming survive across an intervening word? *Journal of Memory and Language, 60*, 36–49.

Forster, K. I., & Chambers, S. M. (1973). Lexical access and naming time. *Journal of Verbal Learning and Verbal Behavior, 12*, 627–635.

Forster, K. I., & Davis, C. (1984). Repetition priming and frequency attenuation in lexical access. *Journal of Experimental Psychology: Learning, Memory, and Cognition, 10*, 680.

Forster, K. I., Davis, C., Schoknecht, C., & Carter, R. (1987). Masked priming with graphemically related forms: Repetition or partial activation? *The Quarterly Journal of Experimental Psychology, 39*, 211–251.

Fowler, C. A., Napps, S. E., & Feldman, L. (1985). Relations among regular and irregular morphologically related words in the lexicon as revealed by repetition priming. *Memory & Cognition, 13*, 241–255.

Frauenfelder, U. H., & Schreuder, R. (1992). Constraining psycholinguistic models of morphological processing and representation: The role of productivity. In G. Booij & J. van Marle (Eds.), *Yearbook of morphology 1991* (pp. 165–183). Dordrecht, Netherlands: Springer.

Friederici, A. D. (1995). The time course of syntactic activation during language processing: A model based on neuropsychological and neurophysiological data. *Brain and Language, 50*, 259–281.

Frost, R., Deutsch, A., & Forster, K. I. (2000a). Decomposing morphologically complex words in a nonlinear morphology. *Journal of Experimental Psychology: Learning, Memory, and Cognition, 26*(3), 751–765.

Frost, R., Deutsch, A., Gilboa, O., Tannenbaum, M., & Marslen-Wilson, W. (2000b). Morphological priming: Dissociation of phonological, semantic, and morphological factors. *Memory & Cognition, 28*, 1277–1288.

Frost, R., Forster, K. I., & Deutsch, A. (1997). What can we learn from the morphology of Hebrew? A masked-priming investigation of morphological representation. *Journal of Experimental Psychology: Learning, Memory, and Cognition, 23*, 829.

Garnsey, S. M. (1993). Event-related brain potentials in the study of language: An introduction. *Language & Cognitive Processes, 8*, 337–356.

Giraudo, H., & Grainger, J. (2000). Effects of prime word frequency and cumulative root frequency in masked morphological priming. *Language & Cognitive Processes, 15*, 421–444.

Giraudo, H., & Grainger, J. (2001). Priming complex words: Evidence for supralexical representation of morphology. *Psychonomic Bulletin & Review, 8*, 127–131.

Gold, B. T., & Rastle, K. (2007). Neural correlates of morphological decomposition during visual word recognition. *Journal of Cognitive Neuroscience, 19*, 1983–1993.

Grainger, J., Colé, P., & Segui, J. (1991). Masked morphological priming in visual word recognition. *Journal of Memory and Language, 30*, 370–384.

Grainger, J., Holcomb, P., Cornelissen, P., & Hansen, P. (2010). Neural constraints on a functional architecture for word recognition. In P. Cornelissen, P. Hansen, M. Kringelbach & K. Pugh (Eds.), *The neural basis of reading* (pp. 3–32). Oxford: Oxford University Press.

Grainger, J., & Holcomb, P. J. (2009). Watching the word go by: On the time-course of component processes in visual word recognition. *Lang & Ling Compass, 3*, 128–156.

Grainger, J., & Jacobs, A. M. (1999). Temporal integration of information in orthographic priming. *Visual Cognition, 6*, 461–492.

Grainger, J., Kiyonaga, K., & Holcomb, P. J. (2006). The time course of orthographic and phonological code activation. *Psychological Science, 17*, 1021–1026.

Grainger, J., & Segui, J. (1990). Neighborhood frequency effects in visual word recognition: A comparison of lexical decision and masked identification latencies. *Attention, Perception, & Psychophysics, 47*, 191–198.

Grainger, J., & Ziegler, J. (2011). A dual-route approach to orthographic processing. *Frontiers in Psychology, 2*, 54.

Hagoort, P., Brown, C., & Groothusen, J. (1993). The syntactic positive shift (SPS) as an ERP measure of syntactic processing. *Language & Cognitive Processes, 8*, 439–483.

Halle, M., & Marantz, A. (1994). Some key features of distributed morphology. *MIT Working Papers in Linguistics, 21*, 88.

Harris, T., Wexler, K., & Holcomb, P. (2000). An ERP investigation of binding and conference. *Brain and Language, 75*, 313–346.

Holcomb, P. J., & Grainger, J. (2006). On the time course of visual word recognition: An event-related potential investigation using masked repetition priming. *Journal of Cognitive Neuroscience, 18*, 1631–1643.

Jaeger, J. J., Lockwood, A. H., Kemmerer, D. L., Van Valin Jr, R. D., Murphy, B. W., & Khalak, H. G. (1996). A positron emission tomographic study of regular and irregular verb morphology in English. *Language, 72*, 451–497.

Kaan, E., Harris, A., Gibson, E., & Holcomb, P. (2000). The P600 as an index of syntactic integration difficulty. *Language & Cognitive Processes, 15*, 159–201.

Kaan, E., & Swaab, T. Y. (2003). Repair, revision, and complexity in syntactic analysis: An electrophysiological differentiation. *Journal of Cognitive Neuroscience, 15*, 98–110.

Kempley, S. T., & Morton, J. (1982). The effects of priming with regularly and irregularly related words in auditory word recognition. *British Journal of Psychology, 73*, 441–454.

Kielar, A., & Joanisse, M. F. (2010). Graded effects of regularity in language revealed by N400 indices of morphological priming. *Journal of Cognitive Neuroscience, 22*, 1373–1398.

Kielar, A., & Joanisse, M. F. (2011). The role of semantic and phonological factors in word recognition: An ERP cross-modal priming study of derivational morphology. *Neuropsychologia, 49*, 161–177.

Kielar, A., Joanisse, M. F., & Hare, M. L. (2008). Priming English past tense verbs: Rules or statistics? *Journal of Memory and Language, 58*, 327–346.

Laudanna, A., Badecker, W., & Caramazza, A. (1989). Priming homographic stems. *Journal of Memory and Language, 28*, 531–546.

Laudanna, A., Badecker, W., & Caramazza, A. (1992). Processing inflectional and derivational morphology. *Journal of Memory and Language, 31*, 333–348.

Laudanna, A., & Burani, C. (1995). Distributional properties of derivational affixes: Implications for processing. In L. B. Feldman (Ed.) *Morphological aspects of language processing* (pp. 345–364). Hillsdale, NJ: Lawrence Erlbaum Associates.

Lavric, A., Clapp, A., & Rastle, K. (2007). ERP evidence of morphological analysis from orthography: A masked priming study. *Journal of Cognitive Neuroscience, 19*, 866–877.

Lavric, A., Elchlepp, H., & Rastle, K. (2012). Tracking hierarchical processing in morphological decomposition with brain potentials. *Journal of Experimental Psychology: Human Perception and Performance, 38*, 811–816.

Lavric, A., Rastle, K., & Clapp, A. (2011). What do fully visible primes and brain potentials reveal about morphological decomposition? *Psychophysiology, 48*, 676–686.

Lehtonen, M., Cunillera, T., Rodríguez-Fornells, A., Hultén, A., Tuomainen, J., & Laine, M. (2007). Recognition of morphologically complex words in Finnish: Evidence from event-related potentials. *Brain Research, 1148*, 123–137.

Lehtonen, M., Monahan, P. J., & Poeppel, D. (2011). Evidence for early morphological decomposition: Combining masked priming with magnetoencephalography. *Journal of Cognitive Neuroscience, 23*, 3366–3379.

Levelt, W. J. (1993). *Speaking: From intention to articulation* (Vol. 1). Cambridge, MA: MIT Press.

Lewis, G., Solomyak, O., & Marantz, A. (2011). The neural basis of obligatory decomposition of suffixed words. *Brain and Language, 118*, 118–127.

Longtin, C.-M., & Meunier, F. (2005). Morphological decomposition in early visual word processing. *Journal of Memory and Language, 53*, 26–41.

Longtin, C.-M., Segui, J., & Hallé, P. A. (2003). Morphological priming without morphological relationship. *Language & Cognitive Processes, 18*, 313–334.
Luck, S. J. (2014). *An introduction to the event-related potential technique*. Cambridge, MA: MIT Press.
Manelis, L., & Tharp, D. A. (1977). The processing of affixed words. *Memory & Cognition, 5*, 690–695.
Marantz, A. (2013). No escape from morphemes in morphological processing. *Language & Cognitive Processes, 28*, 905–916.
Marslen-Wilson, W., Tyler, L. K., Waksler, R., & Older, L. (1994). Morphology and meaning in the English mental lexicon. *Psychological Review, 101*, 3–33.
McCormick, S. F., Rastle, K., & Davis, M. H. (2008). Is there a 'fete'in 'fetish'? Effects of orthographic opacity on morpho-orthographic segmentation in visual word recognition. *Journal of Memory and Language, 58*, 307–326.
McKinnon, R. (1996). Constraints on movement phenomena in sentence processing: Evidence from event-related brain potentials. *Language & Cognitive Processes, 11*, 495–524.
Meunier, F., & Marslen-Wilson, W. (2004). Regularity and irregularity in Frenchverbal inflection. *Language & Cognitive Processes, 19*, 561–580.
Meunier, F., & Segui, J. (1999). Frequency effects in auditory word recognition: The case of suffixed words. *Journal of Memory and Language, 41*, 327–344.
Miceli, G., Turriziani, P., Caltagirone, C., Capasso, R., Tomaiuolo, F., & Caramazza, A. (2002). The neural correlates of grammatical gender: An fMRI investigation. *Journal of Cognitive Neuroscience, 14*, 618–628.
Morris, J., Frank, T., Grainger, J., & Holcomb, P. J. (2007). Semantic transparency and masked morphological priming: An ERP investigation. *Psychophysiology, 44*, 506–521.
Morris, J., Grainger, J., & Holcomb, P. J. (2008). An electrophysiological investigation of early effects of masked morphological priming. *Language & Cognitive Processes, 23*, 1021–1056.
Morris, J., Grainger, J., & Holcomb, P. J. (2013). Tracking the consequences of morpho-orthographic decomposition using ERPs. *Brain Research, 1529*, 92–104.
Morris, J., & Holcomb, P. J. (2005). Event-related potentials to violations of inflectional verb morphology in English. *Cognitive Brain Research, 25*, 963–981.
Morris, J., Porter, J. H., Grainger, J., & Holcomb, P. J. (2011). Effects of lexical status and morphological complexity in masked priming: An ERP study. *Language & Cognitive Processes, 26*, 558–599.
Morris, J., & Stockall, L. (2012). Early, equivalent ERP masked priming effects for regular and irregular morphology. *Brain and Language, 123*, 81–93.
Münte, T. F., Say, T., Clahsen, H., Schiltz, K., & Kutas, M. (1999). Decomposition of morphologically complex words in English: Evidence from event-related brain potentials. *Cognitive Brain Research, 7*, 241–253.
Murrell, G. A., & Morton, J. (1974). Word recognition and morphemic structure. *Journal of Experimental Psychology, 102*, 963.
Napps, S. E. (1989). Morphemic relationships in the lexicon: Are they distinct from semantic and formal relationships? *Memory & Cognition, 17*, 729–739.
Neville, H., Nicol, J. L., Barss, A., Forster, K. I., & Garrett, M. F. (1991). Syntactically based sentence processing classes: Evidence from event-related brain potentials. *Journal of Cognitive Neuroscience, 3*, 151–165.
Orsolini, M., & Marslen-Wilson, W. (1997). Universals in morphological representation: Evidence from Italian. *Language & Cognitive Processes, 12*, 1–47.
Osterhout, L. (1997). On the brain response to syntactic anomalies: Manipulations of word position and word class reveal individual differences. *Brain and Language, 59*, 494–522.
Osterhout, L., & Holcomb, P. J. (1992). Event-related brain potentials elicited by syntactic anomaly. *Journal of Memory and Language, 31*, 785–806.
Osterhout, L., Holcomb, P. J., & Swinney, D. A. (1994). Brain potentials elicited by garden-path sentences: Evidence of the application of verb information during parsing. *Journal of Experimental Psychology: Learning, Memory, and Cognition, 20*, 786–803.

Osterhout, L., & Mobley, L. A. (1995). Event-related brain potentials elicited by failure to agree. *Journal of Memory and Language, 34*, 739–773.

Pastizzo, M. J., & Feldman, L. B. (2002). Discrepancies between orthographic and unrelated baselines in masked priming undermine a decompositional account of morphological facilitation. *Journal of Experimental Psychology: Learning, Memory, and Cognition, 28*, 244–249.

Pastizzo, M. J., & Feldman, L. B. (2004). Morphological processing: A comparison between free and bound stem facilitation. *Brain and Language, 90*, 31–39.

Penke, M., Weyerts, H., Gross, M., Zander, E., Münte, T. F., & Clahsen, H. (1997). How the brain processes complex words: An event-related potential study of German verb inflections. *Cognitive Brain Research, 6*, 37–52.

Pinker, S. (1991). Rules of language. *Science, 253*, 530–535.

Pinker, S. (1997). Words and rules in the human brain. *Nature, 387*, 547.

Pinker, S., & Ullman, M. T. (2002). The past and future of the past tense. *Trends in Cognitive Sciences, 6*, 456–463.

Plunkett, K., & Juola, P. (1999). A connectionist model of English past tense and plural morphology. *Cognitive Science, 23*, 463–490.

Plunkett, K., & Marchman, V. (1993). From rote learning to system building: Acquiring verb morphology in children and connectionist nets. *Cognition, 48*, 21–69.

Prasada, S., & Pinker, S. (1993). Generalisation of regular and irregular morphological patterns. *Language & Cognitive Processes, 8*, 1–56.

Rastle, K., & Davis, M. H. (2008). Morphological decomposition based on the analysis of orthography. *Language & Cognitive Processes, 23*, 942–971.

Rastle, K., Davis, M. H., Marslen-Wilson, W. D., & Tyler, L. K. (2000). Morphological and semantic effects in visual word recognition: A time-course study. *Language & Cognitive Processes, 15*, 507–537.

Rastle, K., Davis, M. H., & New, B. (2004). The broth in my brother's brothel: Morpho-orthographic segmentation in visual word recognition. *Psychonomic Bulletin & Review, 11*, 1090–1098.

Raveh, M. (2002). The contribution of frequency and semantic similarity to morphological processing. *Brain and Language, 81*, 312–325.

Raveh, M., & Rueckl, J. G. (2000). Equivalent effects of inflected and derived primes: Long-term morphological priming in fragment completion and lexical decision. *Journal of Memory and Language, 42*, 103–119.

Rayner, K., White, S. J., Johnson, R. L., & Liversedge, S. P. (2006). Raeding wrods with jubmled lettres there is a cost. *Psychological Science, 17*, 192–193.

Rodriguez-Fornells, A., Clahsen, H., Lleó, C., Zaake, W., & Münte, T. F. (2001). Event-related brain responses to morphological violations in Catalan. *Cognitive Brain Research, 11*, 47–58.

Rodriguez-Fornells, A., Münte, T. F., & Clahsen, H. (2002). Morphological priming in Spanish verb forms: An ERP repetition priming study. *Journal of Cognitive Neuroscience, 14*, 443–454.

Rösler, F., Pütz, P., Friederici, A., & Hahne, A. (1993). Event-related brain potentials while encountering semantic and syntactic constraint violations. *Journal of Cognitive Neuroscience, 5*, 345–362.

Rubin, G. S., Becker, C. A., & Freeman, R. H. (1979). Morphological structure and its effect on visual word recognition. *Journal of Verbal Learning and Verbal Behavior, 18*, 757–767.

Rueckl, J. G., Mikolinski, M., Raveh, M., Miner, C. S., & Mars, F. (1997). Morphological priming, fragment completion, and connectionist networks. *Journal of Memory and Language, 36*, 382–405.

Rueckl, J. G., & Raveh, M. (1999). The influence of morphological regularities on the dynamics of a connectionist network. *Brain and Language, 68*, 110–117.

Rumelhart, D. E., & McClelland, J. L. (1986). On learning the past tenses of English verbs. In J. L. McClelland & D. E. Rumelhart (Eds.), *Parallel distributed processing, Psychological and biological models* (Vol. 2, pp. 216–271). Cambridge, MA: MIT Press.

Schreuder, R., & Baayen, R. H. (1994). Prefix stripping re-revisited. *Journal of Memory and Language, 33*, 357.

Schreuder, R., & Baayen, R. H. (1995). Modeling morphological processing. In L. B. Feldman (Ed.), *Morphological aspects of language processing* (pp. 131–154). Mahwah, NJ: Lawrence Erlbaum Associates Inc..

Schreuder, R., & Baayen, R. H. (1997). How complex simplex words can be. *Journal of Memory and Language, 37*, 118–139.

Seidenberg, M. S., & Gonnerman, L. M. (2000). Explaining derivational morphology as the convergence of codes. *Trends in Cognitive Sciences, 4*, 353–361.

Solomyak, O., & Marantz, A. (2009). Lexical access in early stages of visual word processing: A single-trial correlational MEG study of heteronym recognition. *Brain and Language, 108*, 191–196.

Solomyak, O., & Marantz, A. (2010). Evidence for early morphological decomposition in visual word recognition. *Journal of Cognitive Neuroscience, 22*, 2042–2057.

Sonnenstuhl, I., Eisenbeiss, S., & Clahsen, H. (1999). Morphological priming in the German mental lexicon. *Cognition, 72*, 203–236.

Stanners, R. F., Neiser, J. J., Hernon, W. P., & Hall, R. (1979). Memory representation for morphologically related words. *Journal of Verbal Learning and Verbal Behavior, 18*, 399–412.

Stockall, L., & Marantz, A. (2006). A single route, full decomposition model of morphological complexity: MEG evidence. *The Mental Lexicon, 1*, 85–123.

Taft, M. (1979a). Lexical access-via an orthographic code: The basic orthographic syllabic structure (BOSS). *Journal of Verbal Learning and Verbal Behavior, 18*, 21–39.

Taft, M. (1979b). Recognition of affixed words and the word frequency effect. *Memory & Cognition, 7*, 263–272.

Taft, M. (1988). A morphological-decomposition model of lexical representation. *Linguistics, 26*, 657–668.

Taft, M. (1994). Interactive-activation as a framework for understanding morphological processing. *Language & Cognitive Processes, 9*, 271–294.

Taft, M., & Forster, K. I. (1975). Lexical storage and retrieval of prefixed words. *Journal of Verbal Learning and Verbal Behavior, 14*, 638–647.

Tyler, L. K., Marslen-Wilson, W. D., & Stamatakis, E. A. (2005b). Differentiating lexical form, meaning, and structure in the neural language system. *Proceedings of the National Academy of Sciences of the United States of America, 102*, 8375–8380.

Tyler, L. K., Stamatakis, E. A., Post, B., Randall, B., & Marslen-Wilson, W. (2005a). Temporal and frontal systems in speech comprehension: An fMRI study of past tense processing. *Neuropsychologia, 43*, 1963–1974.

Ullman, M. T. (2001). A neurocognitive perspective on language: The declarative/procedural model. *Nature Reviews Neuroscience, 2*, 717–726.

Ullman, M. T. (2004). Contributions of memory circuits to language: The declarative/procedural model. *Cognition, 92*, 231–270.

Ullman, M. T., Corkin, S., Coppola, M., Hickok, G., Growdon, J. H., Koroshetz, W. J., et al. (1997). A neural dissociation within language: Evidence that the mental dictionary is part of declarative memory, and that grammatical rules are processed by the procedural system. *Journal of Cognitive Neuroscience, 9*, 266–276.

Van Heuven, W. J., Dijkstra, T., Grainger, J., & Schriefers, H. (2001). Shared neighborhood effects in masked orthographic priming. *Psychonomic Bulletin & Review, 8*, 96–101.

Vannest, J., Bertram, R., Järvikivi, J., & Niemi, J. (2002). Counterintuitive cross-linguistic differences: More morphological computation in English than in Finnish. *Journal of Psycholinguistic Research, 31*, 83–106.

Vannest, J., & Boland, J. E. (1999). Lexical morphology and lexical access. *Brain and Language, 68*, 324–332.

Vannest, J., Polk, T. A., & Lewis, R. L. (2005). Dual-route processing of complex words: New fMRI evidence from derivational suffixation. *Cognitive, Affective, & Behavioral Neuroscience, 5*, 67–76.

De Vincenzi, M., Job, R., Di Matteo, R., Angrilli, A., Penolazzi, B., Ciccarelli, L., et al. (2003). Differences in the perception and time course of syntactic and semantic violations. *Brain and Language, 85*, 280–296.

Voga, M., & Grainger, J. (2004). Masked morphological priming with varying levels of form overlap: Evidence from Greek verbs. *Current Psychology Letters: Behaviour, Brain & Cognition*, 13(2).

Weyerts, H., Münte, T. F., Smid, H. G., & Heinze, H.-J. (1996). Mental representations of morphologically complex words: An event-related potential study with adult humans. *Neuroscience Letters, 206*, 125–128.

Zweig, E., & Pylkkänen, L. (2009). A visual M170 effect of morphological complexity. *Language & Cognitive Processes, 24*, 412–439.

Zwitserlood, P., Bolwiender, A., & Drews, E. (2005). Priming morphologically complex verbs by sentence contexts: Effects of semantic transparency and ambiguity. *Language & Cognitive Processes, 20*, 395–415.

Chapter 4
Brain Oscillations, Semantic Processing, and Episodic Memory

Marie-Christin Fellner and Simon Hanslmayr

4.1 Introduction

In this chapter, we try to link empirical neurophysiological findings and concepts from two different disciplines, namely *Semantic processing*, a sub-discipline of linguistics, and *Long-term memory*, a sub-discipline of psychology. We use semantic processing to refer to any sort of cognitive processing which focuses on the meaning of a sensory stimulus (word, picture, or sound). By long-term memory we mean episodic memory, which refers to memories that have a unique temporal and spatial context (Tulving 1972). Importantly, semantic processing and episodic memory have a long-standing tradition in psychology and the combination of these two disciplines led to several key findings and strongly influenced memory models and frameworks (e.g., Tulving 1972; Craik and Lockhart 1972). In keeping with the theme of this book geared towards neural mechanisms, we focus here on a special marker of neural activity, namely brain oscillations, which will provide the glue with which we link the two different disciplines.[1]

In this chapter we focus on studies utilizing brain oscillations to address the question of how local and global neural assemblies interact by means of synchronization and desynchronization during semantic processing and memory encoding.

---

[1] This chapter is written by two psychologists, which may excuse our imprecise usage of linguistic terms.

M.-C. Fellner (✉)
Department of Neuropsychology, Institute of Cognitive Neuroscience,
Ruhr University, Bochum, Germany
e-mail: marie-christin.fellner@ruhr-uni-bochum.de

S. Hanslmayr
School of Psychology, University of Birmingham, Birmingham B15 2TT, UK
e-mail: s.hanslmayr@bham.ac.uk

M. Mody (ed.), *Neural Mechanisms of Language*, Innovations in Cognitive Neuroscience, DOI 10.1007/978-1-4939-7325-5_4

There is still no conclusive answer on how the brain carries out these tasks, but brain oscillations might contribute an important piece to the solution of this puzzle. We will end this chapter with a mechanistic, and perhaps novel, view of how desynchronized and synchronized brain oscillatory activity may guide semantic processing and memory. This idea is a refined version of the previously published *Information* via *Desynchronization Hypothesis* (IDH; Hanslmayr et al. 2012), a hypothesis that still requires further experimental testing, but could quite possibly stimulate further research.

4.2 Brain Oscillations: Basic Concepts

It is beyond the scope of this chapter to give an exhaustive account of brain oscillations (see Buzsaki 2006); instead, we present a brief overview to familiarize the reader with concepts necessary for understanding the content later on. We refer the interested reader to the specific literature cited for in-depth information.

Brain oscillations refer to regular fluctuations of summated electric activity of large numbers of neurons (>10,000), which are recorded by the local field potential (LFP). Thereby, brain oscillations index the summated graded excitatory or inhibitory postsynaptic potentials, which are picked up by EEG (electroencephalography) or MEG (magnetencephalography) sensors (see Hämäläinen et al. 1993; or Schomer & da Silva, 2012, for in-depth reviews). Although some brain oscillations are clearly visible in the healthy human brain with the naked eye, like alpha oscillations which oscillate at around 10 Hz, other oscillations are only revealed after applying time frequency transformation to the raw signal (see Cohen 2014). Fig. 4.1a depicts how a raw EEG signal may be decomposed into a time-frequency representation, revealing that the raw signal is composed of several different oscillations. Brain oscillations are unambiguously defined by three physical properties: (1) frequency (Fig. 4.1b), (2) amplitude (Fig. 4.1c), and (3) phase (Fig. 4.1d).

4.2.1 *Frequency*

Different brain networks are hypothesized to oscillate in different frequencies, with small networks oscillating at fast frequencies (>40 Hz) and large networks oscillating at slower frequencies (<20 Hz) (Von Stein and Sarnthein 2000; Buzsáki and Draguhn 2004). This anatomical property is also reflected in the 1/F power ratio of EEG signals, with exception of the alpha band, suggesting that slower oscillations recruit large pools of neurons and higher oscillations recruit smaller pools of neurons (Fig. 4.1b). This assumption is also in line with the notion that slow oscillations represent global networks of higher hierarchy, which gate faster—more local—oscillations in a top-down manner (Lakatos et al. 2005). Note that small and large networks do not necessarily refer to the physical extent of these networks in topographic

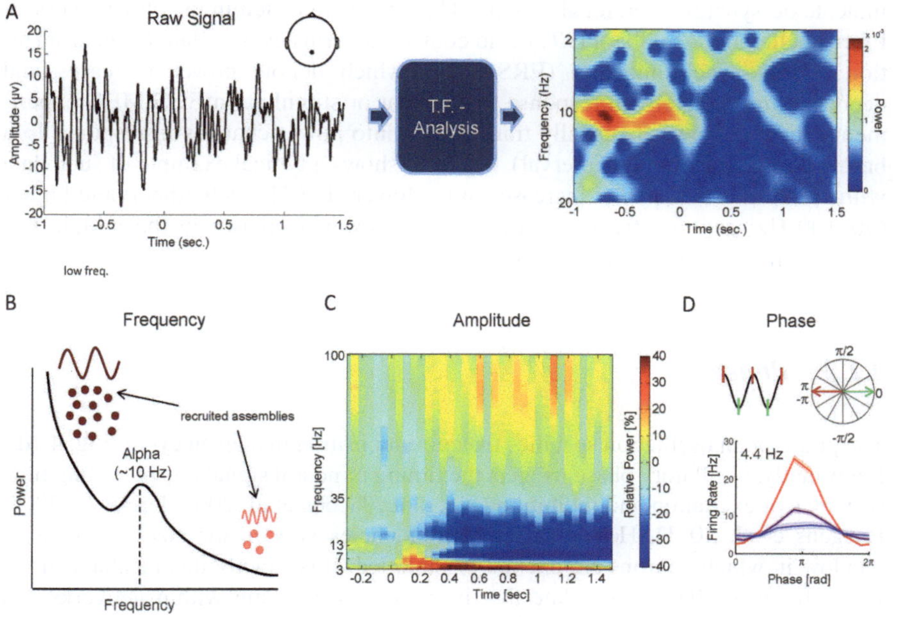

Fig. 4.1 Brain oscillations and their parameters. (**a**) An example of a raw signal as recorded with a parietal EEG electrode is shown on the left. A stimulus was presented at time 0. The plot on the right shows the results of a time-frequency analysis in which the amplitude is depicted for each time-point (x-axis) and frequency band (y-axis). (**b**) A schematic of a typical EEG power spectrum is shown, with frequency on the x-axis and power on the y-axis. The inverse relationship between the size of neural assemblies and power is depicted. Note the peak at the alpha frequency which violates the 1/F relation between power and frequency. (**c**) A typical time-frequency plot showing event-related power increases (*hot colours*) and decreases (*cold colours*) during processing of verbal information. Note the power increases in theta (3–7 Hz) and gamma (35–100 Hz) and the power decreases in alpha (8–12 Hz) and beta (13–35 Hz). (**d**) The relationship between EEG phase (top) and firing rates (bottom) is shown. [**a** is reprinted with permission from Hanslmayr et al. (2011a); **c** is modified and reprinted with permission from Hanslmayr et al. (2012); and **d** modified and reprinted with permission from Jacobs et al. (2007)

space (the Euclidian space), but rather refer to the number of connections involved in a network. Thus, a fronto-parietal network with monosynaptic connections could be a small network, despite covering a distance of several centimeters.

4.2.2 Amplitude/Power

The amplitude, also referred to as power (both terms are used synonymously here), of an oscillation is assumed to reflect the degree of synchrony between inhibitory or excitatory postsynaptic potentials of an underlying neural assembly. Thereby, amplitude increases indicate increased local synchrony and amplitude decreases indicate de-synchronized local activity. This idea is reflected in the classic work of Pfurtscheller and Aranibar (1977) who coined the terms event-related synchronization and de-synchronization (ERS/ERD), which denote power increases and decreases, respectively, in response to an event or stimulus. In EEG/MEG experiments absolute power is usually transformed into power changes in response to a baseline (e.g., prestimulus interval). Fig. 4.1c shows a typical example of such data with stimulus driven power increases in the lower (1–8 Hz: delta/theta) and higher (40–100 Hz; gamma) frequency ranges, and power decreases in the middle frequency ranges (8–35 Hz; alpha/beta).

4.2.3 Phase

The phase of an oscillation specifies the current position in a given cycle (Fig. 4.1d). Importantly, oscillatory phase triggers the timing of neural signals by providing time windows of excitation and inhibition (Fig. 4.1c; Jacobs et al. 2007; Lee et al. 2005; Haegens et al. 2011). Hence, different frequencies vary in the size of the time-window in which neurons are excited or inhibited. It is exactly this mechanism via which brain oscillations regulate the timing of neural firing within networks and thereby enable distant cell assemblies to communicate with each other (Canolty et al. 2010; Hanslmayr et al. 2013). Phase-coherence measures are therefore widely used as measures of long-range synchronization to infer neural communication between distant cell assemblies (Bressler 1996; Varela et al. 2001; Fries 2005). For the current chapter, this latter point is of crucial interest as memory and semantic processing are both assumed to rely on coordinated activity in large-scale cortical (and possibly also sub-cortical) networks. Note that phase information is also employed in several other measures which do not necessarily reflect long-range synchrony (e.g., phase-coherence across trials which is not of relevance in this chapter).

4.2.4 Amplitude and Phase as Measures of Local and Global Synchrony

In this chapter we focus on two measures of oscillatory activity: (1) phase-coherence and (2) amplitude. Phase-coherence refers to a measure of synchrony between distant cell assemblies, whereas amplitude refers to a measure of synchrony between local cell assemblies (Fig. 4.2). As we will show below, phase-coherence measures may help to identify distributed cortical networks that are engaged during semantic and memory tasks; measures of local synchronization, on the other hand, may help to identify certain brain regions receiving inputs from and sending outputs to multiple distributed cortical assemblies thus acting as hubs during semantic processing and memory formation (e.g., the inferior prefrontal cortex).

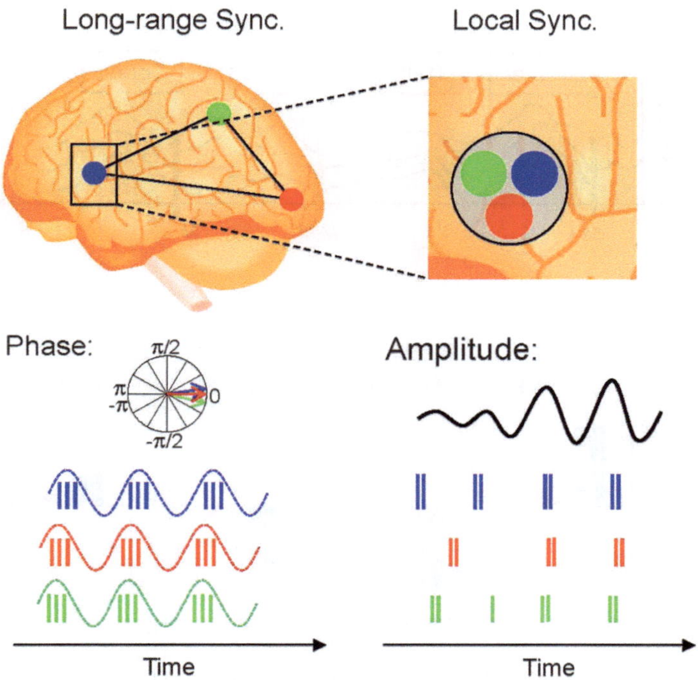

Fig. 4.2 The putative relationship between global (long-range) synchronization between distant neural assemblies and phase-coherence is illustrated in the *left* panel. The right panel depicts the putative relationship between synchronization within local neural assemblies and power (amplitude). *Colored lines* in the *lower panel* refer to the firing rates of the corresponding *colored cell* assemblies in the *upper panels*

4.3 Memory: Basic Concepts

When investigating memory, one first has to consider the type of memory that one is interested in. Any information that is stored over some time and can be consciously reported, without being actively maintained throughout this period, is part of the long-term declarative memory system, in contrast to working memory or conditioned responses, for example. Lesion studies, as exemplified with the famous patient H.M. (Milner et al. 1968), and theoretical considerations showed that declarative memory may be further divided into semantic memory and episodic memory (Tulving 1972). Whereas semantic memory stores concepts and facts, episodic memory is defined by storing the context and the when and where of certain events. So, for example, the information that most people drink coffee for breakfast is part of semantic memory, whereas remembering last Sunday's breakfast, the smell and taste of the food and the conversations you had requires episodic memory.

In the laboratory the neural correlates of episodic memory encoding are investigated with the so-called *subsequent memory paradigm*. Such an experiment

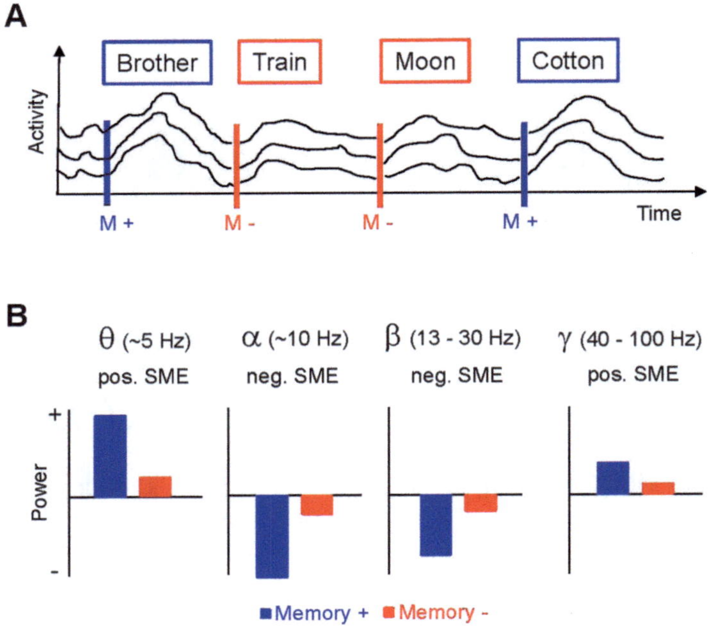

Fig. 4.3 Subsequent Memory Effects: (**a**) In typical memory encoding experiments the subsequent memory paradigm is employed. Activity during encoding is contrasted for items that are later successfully remembered (M+) and items that are later forgotten (M−). (**b**) Studies investigating oscillatory power modulations related to encoding often report an increase in relative theta and gamma power (positive SMEs) and a decrease in relative alpha and beta power (negative SMEs) when comparing the remembered items to the forgotten items

typically consists of an encoding phase, in which items are presented, followed by a short distractor task to prevent items from being maintained in working memory. Finally, items presented in the encoding phase should be remembered during the retrieval phase. The neural correlates of successful memory formation are defined by contrasting activity during the encoding phase depending on memory performance at retrieval (Fig. 4.3a), specifically the contrast of remembered vs. forgotten items which is termed Subsequent Memory Effect (SME). Numerous studies using PET, fMRI, EEG, MEG, and intracranial recordings have investigated SMEs (see Hanslmayr & Staudigl, 2014; Paller and Wagner 2002 for reviews). Although the research appears to be extensive and the paradigm somewhat simple, there is much room for variation and there are still a lot of open questions to be addressed by future research, especially with regard to oscillatory correlates of memory formation.

A vast number of studies have shown the importance of regions in the medial temporal lobe and the hippocampus for successful encoding of episodic memories (Davachi 2006; Paller and Wagner 2002). Importantly, however, the hippocampus is not the only region supporting memory formation; instead, a distributed network of several other cortical areas is also involved in successful memory encoding depending on task demands and the material to be encoded (Kim 2011; Paller and Wagner 2002; Rugg et al. 2008). The left inferior frontal gyrus, for example, is reliably involved in successful memory formation (Blumenfeld and Ranganath 2007; Paller and Wagner 2002; Wagner et al. 1998) especially during encoding of verbal information (Kim 2011). Several other studies investigating the brain oscillatory correlates of successful memory formation have reported a variety of effects (see Fig. 4.3b). Many of these studies report increases (i.e., positive SMEs) in oscillatory theta and gamma power during successful memory formation (Nyhus and Curran 2010). Importantly, decreases (i.e., negative SMEs) in alpha/beta power are also usually observed (Hanslmayr et al. 2012) as well as increases in phase synchronization (Fell and Axmacher 2011). Factors influencing those SMEs are most likely the material used in the experiment, and the way information is processed during the encoding phase and retrieved later (Hanslmayr & Staudigl, 2014).

The phenomenon that later memory is crucially shaped by how information is processed at encoding was first formalized in the famous levels of processing framework (Craik and Lockhart 1972). So-called deep processing during encoding leads to superior later memory performance than shallow encoding. For example, judging whether random words denote something animate leads to better subsequent memory than judging the same words for alphabetical order (e.g., Hanslmayr et al. 2009; Otten and Rugg 2001). The depth of encoding is viewed as a gradient from sensory processing to more meaningful conceptual and semantic processing. However, a problem with this proposal of processing levels and the reason why it is termed a framework and not a comprehensive theory of memory is the lack of a satisfactory definition of "deepness" (Craik 2002). A rule of thumb is that more complex and more elaborate semantic processing of the meaning of a word improves memory performance compared to a less elaborate task (Baddeley 1997). If processing relies on even more conceptual and semantic processing, like organizing items in categories or building stories, it facilitates memory performance even more (Worthen and Hunt 2008).

4.4 Semantic Processing and Memory

The levels of processing framework underline the importance of semantic processing for episodic memory encoding. Memory encoding, according to this framework, can be seen as a byproduct of processing (Craik 2007). Consequently, understanding the neural correlates of semantic processing is important to grasp the processes of memory encoding. In a typical semantic memory encoding task subjects are asked to judge words for animacy or pleasantness, thereby focusing the subjects' attention to the meaning of the words. These tasks that draw on deeper encoding are shown to produce superior memory performance; however, the neural activity associated with it, is less well understood.

Similar to episodic memory, studies investigating semantic processing show widespread effects across the cortex (Binder et al. 2009). The representation of semantic concepts is at least partially overlapping with regions that correspond to perception and action (Pulvermüller 2013). However, in addition to these perceptual and motor areas representing basic sensory features of a certain concept, higher-order (amodal) hub areas integrate these features into the semantic representation (Meteyard et al. 2012). Several cortical areas are known to be involved in semantic tasks and qualify as a semantic hub, e.g., the anterior temporal pole and the left inferior frontal gyrus (Binder and Desai 2011; Martin 2007; Patterson et al. 2007). We propose to focus here on the left inferior frontal gyrus as this region also very reliably shows involvement in memory encoding (Blumenfeld and Ranganath 2007; Paller and Wagner 2002; Wagner et al. 1998).

Several studies have shown that the left IFG is especially involved in memory encoding during semantic processing tasks (Kapur et al. 1994; Otten and Rugg 2001; Otten et al. 2002). Semantic processing studies show that the left IFG is primarily related to controlled semantic retrieval and executive processes (Bookheimer 2002; Noppeney et al. 2004; Thompson-Schill et al. 2005) and combining semantic information (Hagoort 2005). Lesions in this area can lead to rather subtle deficits, like problems in challenging semantic selection tasks (Thompson-Schill et al. 1998) and deficits in understanding abstract words compared to concrete words (Shallice and Cooper 2013). A transcranial magnetic stimulation (TMS) study implicates the left IFG in processing abstract words, caused perhaps by the crucial role of the left IFG in creating meaningful context to understand abstract words (Hoffman et al. 2010). Together, these rather complex semantic operations seem to fit the term "deep" semantic processing used in the levels of processing framework (Craik and Lockhart 1972).

The conceptualization of the semantic system as a distributed cortical network with processing hubs, e.g. the left IFG, is similar to the episodic memory system. Long-range increases in brain oscillatory synchrony are putatively a crucial mechanism for binding information in distant cortical areas into a single representation. Local desynchronization of oscillations, as explained below, might index information processing in local cell assemblies. Therefore analyzing brain oscillatory data is a straightforward way to gain deeper insight into the semantic processing system. Unfortunately, only few studies so far have investigated brain oscillations related to semantic processing (for reviews see Bastiaansen and Hagoort 2006; Hagoort et al. 2004; Weiss and Mueller 2012). Interestingly, some studies localized decreases in beta oscillatory power during semantic processing in the left IFG. Wang et al. (2012) showed that the decreases in beta power in the left IFG index the N400 m effect related to the detection of semantic violations. Also, decreases in beta power in the left IFG were found in a semantic oddball task (Kim and Chung 2008) and word generation task (Singh et al. 2002). These findings will become relevant in the following sections where similar results were obtained in studies looking at episodic memory formation. (As a reminder, we focus on semantic processing; readers interested in the role of different brain oscillations in language more generally may look at work by Canolty et al. (2007) among others.)

4.4.1 Semantic Processing, Memory Formation, and Local Desynchronization

Several studies investigating brain oscillatory correlates of memory encoding show that decreases in beta power index successful memory encoding (Hanslmayr et al. 2012). In fact, decreases in beta power seem to specifically index semantic memory encoding, especially relevant for this chapter on brain oscillations and semantic processing. A first study investigating the effect of varying encoding tasks on brain oscillatory SMEs contrasted a typical shallow alphabetical encoding task with a semantic animacy judgment task (Hanslmayr et al. 2009) (see Fig. 4.4a). Whereas encoding success in the shallow task was related to increases in theta power, encoding in the semantic condition was specifically related to decreases in beta power at left frontal electrode sites. This specific relationship of beta power decreases and successful memory encoding was replicated in an additional study using simultaneous EEG-fMRI (Hanslmayr et al. 2011b). In this study, subjects intentionally encoded word lists and later retrieved those in a free recall task. Again, successful encoding of the words was related to decreases of beta power localized in the left IFG (Fig. 4.4b). Additionally, fMRI BOLD activity revealed an encoding-related increase of activity in the left IFG, which negatively correlated with the beta power memory effect on a trial-by-trial basis (see Fig. 4.4c). This suggests that task related decreases of oscillatory beta power directly index the energy consumption in the left IFG and highlights the active role of power decreases in memory encoding.

The relation of beta power decreases to memory formation in an area crucially involved in semantic processing suggests a vital role for power decreases in semantic memory encoding, albeit it does not imply causality. Therefore, we conducted an additional study using repetitive transcranial magnetic stimulation (rTMS) (Hanslmayr et al. 2014). Using rTMS, it is possible to entrain a focal cortical area in a specific oscillation (Thut and Miniussi 2009), and consequently to probe for causal involvement of brain oscillations in cognitive tasks. In our experiment, subjects encoded word lists while their left IFG was stimulated in three different frequencies (6.8 Hz, 11.7 Hz, 18.7 Hz) and a sham condition without active stimulation (see Fig. 4.5a,b). If beta power decreases in the left IFG play a crucial role in memory encoding, then preventing those decreases by externally increasing beta oscillations by rTMS should selectively decrease memory performance. The behavioral results showed exactly this pattern. Memory performance was significantly decreased for the beta rTMS condition compared to the other stimulation and sham conditions (Fig. 4.5c). Furthermore, only in the beta stimulation condition the EEG of the subjects was entrained to the stimulation frequency, as revealed by an "entrainment echo". This echo was visible in the EEG signal after the actual stimulation had stopped and was localized to the left IFG (Fig. 4.5d). This specific effect of beta frequency stimulation on the left IFG and on memory encoding further underlines the tight link between decreased left IFG beta synchrony and memory encoding.

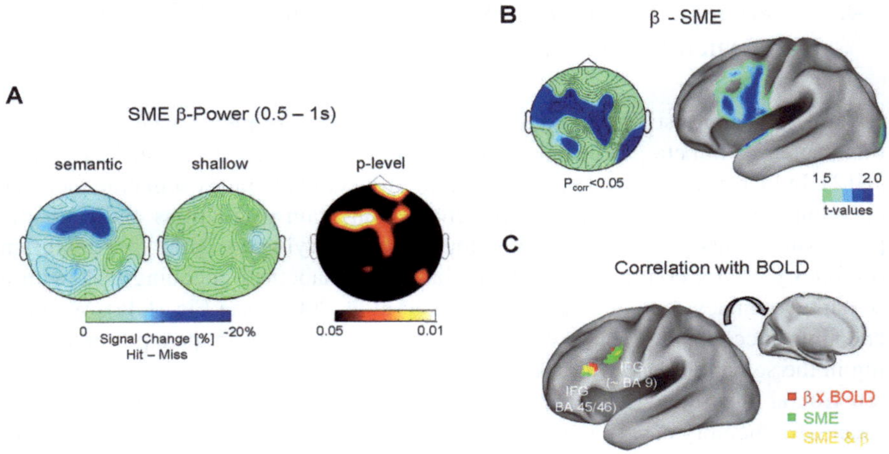

Fig. 4.4 Beta Desynchronization and Semantic Memory Encoding (**a**) Hanslmayr et al. (2009) showed that encoding during a semantic processing task is specifically related to a decrease in left frontal beta power. The topoplots show the encoding related decrease (negative SME) in semantic animacy judgment task and the shallow alphabetical judgment task. The p-level plot shows electrodes site revealing a significant effect. (**b**) An additional simultaneous EEG-fMRI study (Hanslmayr et al., 2011a) replicated the negative SME in the beta frequency band and source localized this effect to left IFG. (**c**) fMRI data revealed an increase in activity in left IFG (SME). The memory-related beta decrease was correlated on a trial-by-trial basis with task-related BOLD activity in left IFG. Figures in **a** reprinted with permission from Hanslmayr et al. (2009); Figures in **b** and **c** reprinted with permission from Hanslmayr et al. (2011b)

Linking these results with findings showing the involvement of the left IFG in complex semantic processing (Bookheimer 2002; Noppeney et al. 2004; Thompson-Schill et al. 2005) and in memory encoding processes (Badre and Wagner 2007; Blumenfeld and Ranganath 2007) further speaks to the importance of semantic processes in episodic memory encoding. Vice versa, it emphasizes the vital role of beta oscillatory desynchronization for memory and semantic processing.

4.4.2 Semantic Processing, Memory Formation, and Global/ Local Synchronization

Oscillatory changes related to memory formation and semantic processing are found in other frequency bands as well. However, these relationships seem not to be as clear-cut as the beta power decreases during semantic memory encoding tasks. The models that describe episodic and semantic memory as cortically distributed information that is retrieved/bound together by cortical hubs suggest the involvement of not only local power synchronization effects but also long-range cortical synchronization (Fell and Axmacher 2011; Von Stein and Sarnthein 2000; Varela et al. 2001).

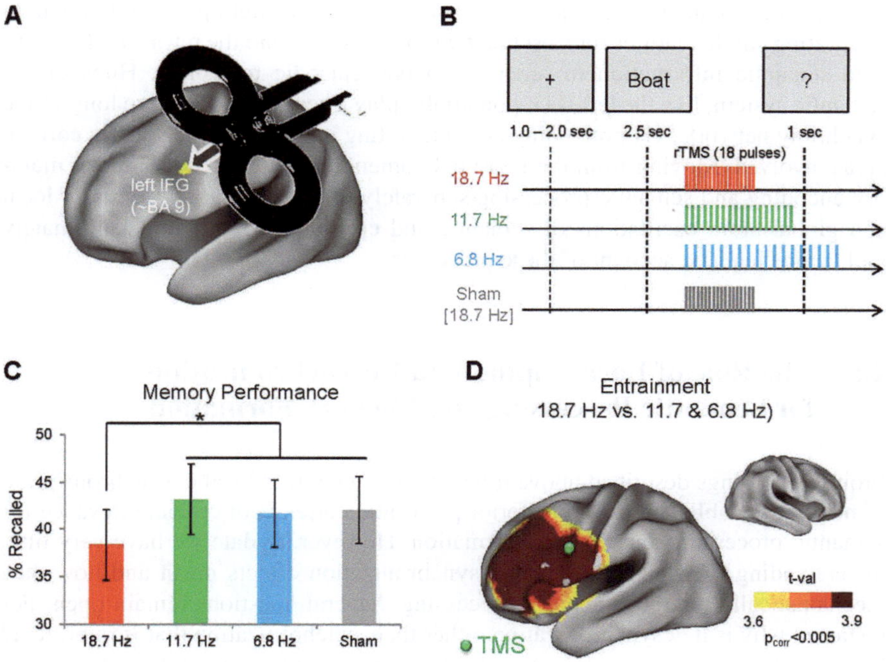

Fig. 4.5 Causal role of beta desynchronization in Memory Encoding (**a**) In order to investigate the causal role of beta power desynchronization in the left IFG during memory formation Hanslmayr et al. (2014) applied rTMS during encoding. (**b**) Stimulation was carried out during the encoding period in the beta (18.7 Hz), alpha (11.7 Hz), theta (6.8 Hz), and a sham condition. (**c**) Memory performance was significantly impaired in the beta stimulation condition compared to all other stimulation conditions. (**d**) Only in the beta stimulation condition there was an entrainment effect evident. After stimulation in the beta frequency, the left frontal cortex showed a prolonged increase (i.e., an echo) in beta synchrony

Indeed, studies looking at memory encoding have reported increases in long-range phase synchronization related to successful memory encoding (Burke et al. 2013; Fell et al. 2001; Fellner et al. 2013; Summerfield and Mangels 2005; Weiss and Rappelsberger 2000). One study that reported increases in alpha and beta long range phase synchronization during memory encoding found these increases especially in a survival encoding task compared to a classical semantic encoding task (animacy judgments) (Fellner et al. 2013). Judging items in this survival task involved more complex, elaborate semantic, emotional, and self-referential processing and therefore presumably involves a more widespread cortical network to integrate task relevant features. This might be reflected in the observed increase in global synchrony. Long-range alpha synchronization has also been proposed as a general mechanism of retrieval of the semantic knowledge system (Klimesch 2012). Moreover, increases in theta long-range synchrony have also been shown to facilitate the encoding of semantically congruent items (Crespo-Garcia et al. 2010).

These task-related increases in long-range synchrony might potentially relate to integrating single features into a semantic representation and the retrieval of distributed semantic information in order to encode episodic memories. Hubs of the semantic system, like the left IFG, potentially play a central role in these long-range synchrony networks via processing and integrating inputs from the various cortical areas involved. Moving from the more phenomenological studies looking at memory encoding and semantic processing separately to more integrated study of local and global brain oscillations of semantic and episodic memory might ultimately yield a mechanistic account of these processes.

4.5 The Role of Local Alpha/Beta Desynchronization for Semantic Processing and Memory Formation

From the findings described above it becomes evident that local desynchronization of neural assemblies in the left inferior prefrontal cortex is of crucial relevance for semantic processing and memory formation. However, to date we have very little understanding of what these local desynchronization effects mean and how they mechanistically relate to memory processing. Several questions remain open. For instance, why is it desynchronization rather than synchronization that is important? Why do these effects occur in specific frequency bands, e.g., alpha/beta oscillations? As pointed out at the beginning of this chapter, we do not have comprehensive answers to these questions at this time but try to offer answers, derived from a recently proposed model (Hanslmayr et al. 2016).

Intuitively, one might think that it is synchronization, rather than desynchronization, that allows local neuron assemblies to communicate and process information and strengthen their synaptic plasticity, which renders the above presented findings even more puzzling. However, intuition might fail here as revealed by the fact that brain states which are associated with very low degrees of information processing, such as coma, epileptic seizures or deep sleep, are usually characterized by very high levels of synchrony. In the normal healthy and awake brain, the transition from synchronized to desynchronized brain states during information processing is ubiquitous (Harris and Thiele 2011) and so dominant that it was one of the first phenomena described in the human EEG (Berger 1933; Pfurtscheller and Aranibar 1977). In order to link the desynchronization in alpha and beta oscillations to memory and semantic processing we recently proposed the Information via Desynchronization Hypothesis (IDH), stating that on a neural level, information is best represented in desynchronized spiking activity. This idea was derived from neuro-computational accounts using information theory (Barlow 1961; Bialek et al. 1991; Schneidman et al. 2011). Information theory states that the information that is conveyed by a certain event (e.g., neural spike) is negatively related to its predictability. In other words, an event which happens at a time where it is highly expected carries less information than an event happening at a time when it is unexpected (Shannon & Weaver, 1963). This is typically illustrated by a coin tossing game. If one tosses a coin, there are two equally probable outcomes and each new toss will give new information, as each toss is independent from the previous coin tosses. Therefore, the information that is obtained in such a coin tossing game is maximal with 1 bit per toss, as calculated with Shannon's Entropy, which is a measure of the richness of information. If the coin would be faked such that each toss produces the same outcome, Shannon's Entropy would be zero, indicating minimal degree of information.

Applying this concept to the firing rates of neurons immediately reveals that the richness of information that is encoded in the firing rates of a neural assembly is negatively related to the synchrony within these firing patterns. This can easily be revealed by a simple simulation, which is depicted in Fig. 4.6a, simulating firing patterns of a neural assembly ($N = 50$) with varying degrees of synchronization. Importantly, the sum of spikes is held constant between the different synchronization conditions (approx. 450). Computing Shannon's Entropy from these differently synchronized firing patterns reveals that the degree of information that is encoded in the assembly decreases as a function of synchronization, with the richest information being encoded in the most desynchronized firing pattern (Fig. 4.6b). This simulation illustrates that the more information needs to be encoded, the more desynchronized the firing of local neural assemblies needs to be. We hypothesize that in human electrophysiological activity such desynchronization demands are reflected by the relative decreases in alpha/beta amplitudes.

Although this framework is based on very simple assumptions it provides a straightforward view as to how desynchronization in local neural assemblies enables hub regions to process semantic information and promote memory storage. However, it might be too simplistic a framework in that it neglects the increases

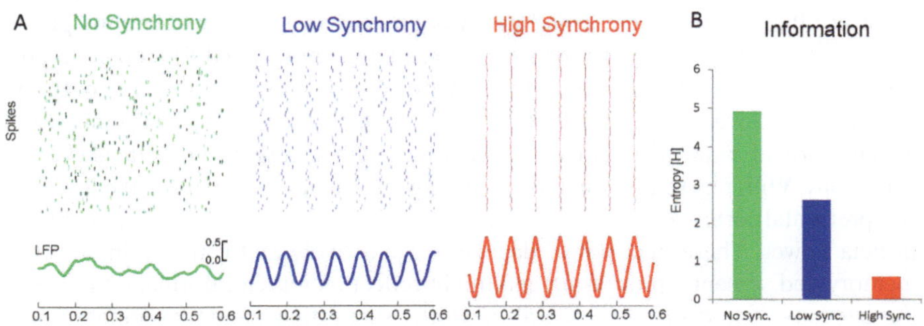

Fig. 4.6 The link between desynchronization and information. (**a**) Firing rates for a population of neurons ($N = 50$) was simulated with either no synchrony (*left*), a low degree of synchrony (*middle*), or a high degree of synchrony (*right*). The total number of spikes in each population was the same. The lower panels plot the corresponding local field potentials (LFP). (**b**) The plot shows information, calculated with Shannon's Entropy, derived from the firing rates of the three neural populations. Figure adapted and reprinted with permission from Hanslmayr et al. (2012)

in local synchrony that accompany the alpha/beta power decreases, especially in the theta and gamma frequency range. In a recent paper, Brittain and Brown (2014) offered an integrative view, highlighting that both synchronized and desynchronized oscillatory activity are important and that it might be the balance between those two states that is critical for information processing. These authors proposed that brain oscillations could be divided into two classes of oscillations, based on whether they promote immutability or mutability. Immutability promoting oscillations are oscillations which are generally high during inactivity and decrease during processing, therefore enhancing information processing capacity in local assemblies (i.e., alpha and beta oscillations). Mutability promoting oscillations are usually very low during inactivity and increase during processing (e.g., theta and gamma). These mutability promoting oscillations selectively synchronize neurons, thus enabling information transfer within an assembly. In a similar way, increases in global synchronization (i.e., phase coherence) likely enable the processing of semantic information represented in distant neural assemblies. Crucially, such phase synchronization might promote long-term potentiation which is a key mechanism for memory formation (Fell and Axmacher 2011). Albeit being quite speculative at the moment, these concepts might be useful in illustrating an integrative way of thinking in which both synchronization and desynchronization of neurons are being considered for information processing, as opposed to frameworks which only focus on one of the two.

4.6 Summary and Conclusions

In this chapter, we reviewed recent findings giving insights into the neural mechanisms of memory formation and semantic processing with a special emphasis on brain oscillations. From these findings it is obvious that brain oscillations play a very important role for the formation of episodic memories and for semantic processing in that they enable neural computations to be carried out at the local and global neural levels. Specifically, hub regions which process inputs from several distant cortical brain regions can be identified with increased or decreased local synchrony. With regard to semantic processing and memory formation, the left inferior prefrontal cortex seems to be such a hub region and is characterized by decreases in beta power. These beta power decreases are actively and causally involved in memory and semantic processing and might reflect increased information processing capacity. The distributed neural networks supporting memory formation and semantic processing, on the other hand, might reflect information processing in distant neural assemblies required to bind several features into one coherent semantic—episodic memory representation.

Acknowledgements The work presented here was supported by a grant from the German Research Foundation (DFG HA5622/1-1) awarded to S.H.

References

Baddeley, A. D. (1997). *Human memory: Theory and practice*. East Sussex: Psychology Press.

Badre, D., & Wagner, A. D. (2007). Left ventrolateral prefrontal cortex and the cognitive control of memory. *Neuropsychologia, 45*, 2883–2901.

Barlow, H. B. (1961). The coding of sensory messages. *Current Problems in Animal Behaviour, 331*, 360.

Bastiaansen, M., & Hagoort, P. (2006). Oscillatory neuronal dynamics during language comprehension. *Progress in Brain Research, 159*, 179–196.

Berger, H. (1933). Über das elektrenkephalogramm des menschen. *Archiv für Psychiatrie und Nervenkrankheiten, 99*, 555–574.

Bialek, W., Rieke, F., de van Steveninck, R. R., & Warland, D. (1991). Reading a neural code. *Science, 252*, 36–43.

Binder, J. R., & Desai, R. H. (2011). The neurobiology of semantic memory. *Trends in Cognitive Sciences, 15*, 527–536.

Binder, J. R., Desai, R. H., Graves, W. W., & Conant, L. L. (2009). Where is the semantic system? A critical review and meta-analysis of 120 functional neuroimaging studies. *Cerebral Cortex, 19*, 2767–2796.

Blumenfeld, R. S., & Ranganath, C. (2007). Prefrontal cortex and long-term memory encoding: An integrative review of findings from neuropsychology and neuroimaging. *The Neuroscientist, 13*, 280–291.

Bookheimer, S. (2002). Functional MRI of language: New approaches to understanding the cortical organization of semantic processing. *Annual Review of Neuroscience, 25*, 151–188.

Bressler, S. L. (1996). Interareal synchronization in the visual cortex. *Behavioural Brain Research, 76*, 37–49.

Brittain, J.-S., & Brown, P. (2014). Oscillations and the basal ganglia: Motor control and beyond. *NeuroImage, 85*, 637–647.

Burke, J. F., Zaghloul, K. A., Jacobs, J., Williams, R. B., Sperling, M. R., Sharan, A. D., et al. (2013). Synchronous and asynchronous theta and gamma activity during episodic memory formation. *Journal of Neuroscience, 33*, 292–304.

Buzsaki, G. (2006). *Rhythms of the brain*. Oxford: Oxford University Press.

Buzsáki, G., & Draguhn, A. (2004). Neuronal oscillations in cortical networks. *Science, 304*, 1926–1929.

Canolty, R. T., Ganguly, K., Kennerley, S. W., Cadieu, C. F., Koepsell, K., Wallis, J. D., et al. (2010). Oscillatory phase coupling coordinates anatomically dispersed functional cell assemblies. *Proceedings of the National Academy of Sciences, 107*, 17356–17361.

Canolty, R. T., Soltani, M., Dalal, S. S., Edwards, E., Dronkers, N. F., Nagarajan, S. S., et al. (2007). Spatiotemporal dynamics of word processing in the human brain. *Frontiers in Neuroscience, 1*, 14.

Cohen, M. X. (2014). *Analyzing neural time series data: Theory and practice*. Cambridge, MA: MIT Press.

Craik, F. I. (2002). Levels of processing: Past, present… and future? *Memory, 10*, 305–318.

Craik, F. I. M. (2007). Encoding: A cognitive perspective. In H. L. Roediger, Y. Dudai, & S. M. Fitzpatrick (Eds.), *Science of memory: Concepts*. New York: Oxford University Press.

Craik, F. I., & Lockhart, R. S. (1972). Levels of processing: A framework for memory research. *Journal of Verbal Learning and Verbal Behavior, 11*, 671–684.

Crespo-Garcia, M., Cantero, J. L., Pomyalov, A., Boccaletti, S., & Atienza, M. (2010). Functional neural networks underlying semantic encoding of associative memories. *NeuroImage, 50*, 1258–1270.

Davachi, L. (2006). Item, context and relational episodic encoding in humans. *Current Opinion in Neurobiology, 16*, 693–700.

Fell, J., & Axmacher, N. (2011). The role of phase synchronization in memory processes. *Nature Reviews Neuroscience, 12*, 105–118.

Fell, J., Klaver, P., Lehnertz, K., Grunwald, T., Schaller, C., Elger, C. E., et al. (2001). Human memory formation is accompanied by rhinal–hippocampal coupling and decoupling. *Nature Neuroscience, 4*, 1259–1264.

Fellner, M.-C., Bäuml, K.-H. T., & Hanslmayr, S. (2013). Brain oscillatory subsequent memory effects differ in power and long-range synchronization between semantic and survival processing. *NeuroImage, 79*, 361–370.

Fries, P. (2005). A mechanism for cognitive dynamics: Neuronal communication through neuronal coherence. *Trends in Cognitive Sciences, 9*, 474–480.

Haegens, S., Nácher, V., Luna, R., Romo, R., & Jensen, O. (2011). α-oscillations in the monkey sensorimotor network influence discrimination performance by rhythmical inhibition of neuronal spiking. *Proceedings of the National Academy of Sciences, 108*, 19377–19382.

Hagoort, P. (2005). On Broca, brain, and binding: A new framework. *Trends in Cognitive Sciences, 9*, 416–423.

Hagoort, P., Hald, L., Bastiaansen, M., & Petersson, K. M. (2004). Integration of word meaning and world knowledge in language comprehension. *Science, 304*, 438–441.

Hämäläinen, M., Hari, R., Ilmoniemi, R. J., Knuutila, J., & Lounasmaa, O. V. (1993). Magnetoencephalography—Theory, instrumentation, and applications to noninvasive studies of the working human brain. *Reviews of Modern Physics, 65*, 413.

Hanslmayr, S., Gross, J., Klimesch, W., & Shapiro, K. L. (2011a). The role of alpha oscillations in temporal attention. *Brain Research Reviews, 67*, 331–343.

Hanslmayr, S., Volberg, G., Wimber, M., Raabe, M., Greenlee, M. W., & Bäuml, K.-H. T. (2011b). The relationship between brain oscillations and BOLD signal during memory formation: A combined EEG–fMRI study. *Journal of Neuroscience, 31*, 15674–15680.

Hanslmayr, S., Matuschek, J., & Fellner, M.-C. (2014). Entrainment of prefrontal beta oscillations induces an endogenous echo and impairs memory formation. *Current Biology, 24*, 904–909.

Hanslmayr, S., Staresina, B. P., & Bowman, H. (2016). Oscillations and episodic memory: Addressing the synchronization/desynchronization conundrum. *Trends in Neurosciences, 39*, 16–25.

Hanslmayr, S., & Staudigl, T. (2014). How brain oscillations form memories—A processing based perspective on oscillatory subsequent memory effects. *NeuroImage, 85*, 648–655.

Hanslmayr, S., Staudigl, T., & Fellner, M.-C. (2012). Oscillatory power decreases and long-term memory: The information via desynchronization hypothesis. *Frontiers in Human Neuroscience, 6*, 74.

Hanslmayr, S., Spitzer, B., & Bäuml, K.-H. (2009). Brain oscillations dissociate between semantic and nonsemantic encoding of episodic memories. *Cerebral Cortex, 19*, 1631–1640.

Hanslmayr, S., Volberg, G., Wimber, M., Dalal, S. S., & Greenlee, M. W. (2013). Prestimulus oscillatory phase at 7 Hz gates cortical information flow and visual perception. *Current Biology, 23*, 2273–2278.

Harris, K. D., & Thiele, A. (2011). Cortical state and attention. *Nature Reviews Neuroscience, 12*, 509–523.

Hoffman, P., Jefferies, E., & Ralph, M. A. L. (2010). Ventrolateral prefrontal cortex plays an executive regulation role in comprehension of abstract words: Convergent neuropsychological and repetitive TMS evidence. *Journal of Neuroscience, 30*, 15450–15456.

Jacobs, J., Kahana, M. J., Ekstrom, A. D., & Fried, I. (2007). Brain oscillations control timing of single-neuron activity in humans. *Journal of Neuroscience, 27*, 3839–3844.

Kapur, S., Craik, F. I., Tulving, E., Wilson, A. A., Houle, S., & Brown, G. M. (1994). Neuroanatomical correlates of encoding in episodic memory: Levels of processing effect. *Proceedings of the National Academy of Sciences, 91*, 2008–2011.

Kim, H. (2011). Neural activity that predicts subsequent memory and forgetting: A meta-analysis of 74 fMRI studies. *NeuroImage, 54*, 2446–2461.

Kim, J. S., & Chung, C. K. (2008). Language lateralization using MEG beta frequency desynchronization during auditory oddball stimulation with one-syllable words. *NeuroImage, 42*, 1499–1507.

Klimesch, W. (2012). Alpha-band oscillations, attention, and controlled access to stored information. *Trends in Cognitive Sciences, 16*, 606–617.

Lakatos, P., Shah, A. S., Knuth, K. H., Ulbert, I., Karmos, G., & Schroeder, C. E. (2005). An oscillatory hierarchy controlling neuronal excitability and stimulus processing in the auditory cortex. *Journal of Neurophysiology, 94*, 1904–1911.

Lee, H., Simpson, G. V., Logothetis, N. K., & Rainer, G. (2005). Phase locking of single neuron activity to theta oscillations during working memory in monkey extrastriate visual cortex. *Neuron, 45*, 147–156.

Martin, A. (2007). The representation of object concepts in the brain. *Annual Review of Psychology, 58*, 25–45.

Meteyard, L., Cuadrado, S. R., Bahrami, B., & Vigliocco, G. (2012). Coming of age: A review of embodiment and the neuroscience of semantics. *Cortex, 48*, 788–804.

Milner, B., Corkin, S., & Teuber, H.-L. (1968). Further analysis of the hippocampal amnesic syndrome: 14-year follow-up study of HM. *Neuropsychologia, 6*, 215–234.

Noppeney, U., Phillips, J., & Price, C. (2004). The neural areas that control the retrieval and selection of semantics. *Neuropsychologia, 42*, 1269–1280.

Nyhus, E., & Curran, T. (2010). Functional role of gamma and theta oscillations in episodic memory. *Neuroscience & Biobehavioral Reviews, 34*, 1023–1035.

Otten, L. J., Henson, R. N., & Rugg, M. D. (2002). State-related and item-related neural correlates of successful memory encoding. *Nature Neuroscience, 5*, 1339–1344.

Otten, L. J., & Rugg, M. D. (2001). Task-dependency of the neural correlates of episodic encoding as measured by fMRI. *Cerebral Cortex, 11*, 1150–1160.

Paller, K. A., & Wagner, A. D. (2002). Observing the transformation of experience into memory. *Trends in Cognitive Sciences, 6*, 93–102.

Patterson, K., Nestor, P. J., & Rogers, T. T. (2007). Where do you know what you know? The representation of semantic knowledge in the human brain. *Nature Reviews Neuroscience, 8*, 976–987.

Pfurtscheller, G., & Aranibar, A. (1977). Event-related cortical desynchronization detected by power measurements of scalp EEG. *Electroencephalography and Clinical Neurophysiology, 42*, 817–826.

Pulvermüller, F. (2013). How neurons make meaning: Brain mechanisms for embodied and abstract-symbolic semantics. *Trends in Cognitive Sciences, 17*, 458–470.

Rugg, M. D., Johnson, J. D., Park, H., & Uncapher, M. R. (2008). Encoding-retrieval overlap in human episodic memory: A functional neuroimaging perspective. *Progress in Brain Research, 169*, 339–352.

Schneidman, E., Puchalla, J. L., Segev, R., Harris, R. A., Bialek, W., & Berry, M. J. (2011). Synergy from silence in a combinatorial neural code. *Journal of Neuroscience, 31*, 15732–15741.

Schomer, D. L., & Da Silva, F. L. (2012). *Niedermeyer's electroencephalography: Basic principles, clinical applications, and related fields*. Philadelphia, PA: Lippincott Williams & Wilkins.

Shallice, T., & Cooper, R. P. (2013). Is there a semantic system for abstract words? *Frontiers in Human Neuroscience, 7*, 175.

Shannon, C. E., & Weaver, W. (1963). *The mathematical theory of communication*. Champaign, IL: University of Illinois Press.

Singh, K. D., Barnes, G. R., Hillebrand, A., Forde, E. M., & Williams, A. L. (2002). Task-related changes in cortical synchronization are spatially coincident with the hemodynamic response. *NeuroImage, 16*, 103–114.

Summerfield, C., & Mangels, J. A. (2005). Coherent theta-band EEG activity predicts item-context binding during encoding. *NeuroImage, 24*, 692–703.

Thompson-Schill, S. L., Bedny, M., & Goldberg, R. F. (2005). The frontal lobes and the regulation of mental activity. *Current Opinion in Neurobiology, 15*, 219–224.

Thompson-Schill, S. L., Swick, D., Farah, M. J., D'Esposito, M., Kan, I. P., & Knight, R. T. (1998). Verb generation in patients with focal frontal lesions: A neuropsychological test of neuroimaging findings. *Proceedings of the National Academy of Sciences, 95*, 15855–15860.

Thut, G., & Miniussi, C. (2009). New insights into rhythmic brain activity from TMS–EEG studies. *Trends in Cognitive Sciences, 13*, 182–189.

Tulving, E. (1972). Episodic and semantic memory. In *Organization of memory* (pp. 381–402). New York: Academic Press.

Varela, F., Lachaux, J.-P., Rodriguez, E., & Martinerie, J. (2001). The brainweb: Phase synchronization and large-scale integration. *Nature Reviews Neuroscience, 2*, 229–239.

Von Stein, A., & Sarnthein, J. (2000). Different frequencies for different scales of cortical integration: From local gamma to long range alpha/theta synchronization. *International Journal of Psychophysiology, 38*, 301–313.

Wagner, A. D., Schacter, D. L., Rotte, M., Koutstaal, W., Maril, A., Dale, A. M., et al. (1998). Building memories: Remembering and forgetting of verbal experiences as predicted by brain activity. *Science, 281*, 1188–1191.

Wang, L., et al. (2012). Beta oscillations relate to the N400m during language comprehension. *Human Brain Mapping, 33*, 2898–2912.

Weiss, S., & Mueller, H. M. (2012). "too many betas do not spoil the broth": The role of beta brain oscillations in language processing. *Frontiers in Psychology, 3*, 201.

Weiss, S., & Rappelsberger, P. (2000). Long-range EEG synchronization during word encoding correlates with successful memory performance. *Cognitive Brain Research, 9*, 299–312.

Worthen, J. B., & Hunt, R. R. (2008). 2.10-mnemonics: Underlying processes and practical applications. In J. H. Byrne (Ed.), *Learning and memory: A comprehensive reference* (pp. 145–156). Oxford: Academic Press.

Chapter 5
The Role of Language in Structure-Dependent Cognition

Martin M. Monti

5.1 Introduction

The ability to construct an indefinite number of ideas by combining a finite set of elements in a hierarchically structured sequence is a signal characteristic of human cognition. To illustrate, consider the sentence *The girl who kissed the boy closed the door*. It is immediately clear to any proficient English speaker that the state of affairs described by this sentence is that the girl is doing the closing. This specific interpretation is as effortless as automatic, and if anybody interpreted it any differently it might be sufficient grounds for doubting her proficiency of the English language. Nonetheless, one might wonder why, for example, we do not interpret the noun phrase *the boy* as being the subject of the verb phrase *closed the door*. After all, *the boy* is linearly much more proximal to the verb phrase than is *the girl*. Furthermore, the sentence actually even contains the well-formed fragment [...] *the boy closed the door*, in which, of course, it is the boy doing the closing. Yet, when we consider the full sentence, the relative linear proximity of its component elements does not appear to guide our interpretation. How is it then that we so effortlessly and automatically interpret the sentence above as describing a state of affairs in which a (certain) girl, who just so happens to have given a kiss to a (certain) boy, has closed the door? One explanation, which is perhaps the founding intuition of the modern study of language as a mental phenomenon, is that despite the fact that language is typically manifested as a temporally linear sequence of utterances, in our mind we spontaneously build a rich abstract hierarchical representation of how each discrete element within the sequence relates to every other element. It is the building of these abstract representations that allows us to assign meaning to strings of utterances. Although this ability is most prominently displayed in our use of natural language,

M.M. Monti (✉)
Department of Psychology, UCLA, Los Angeles, CA 90095, USA
e-mail: monti@psych.ucla.edu

M. Mody (ed.), *Neural Mechanisms of Language*, Innovations in Cognitive Neuroscience, DOI 10.1007/978-1-4939-7325-5_5

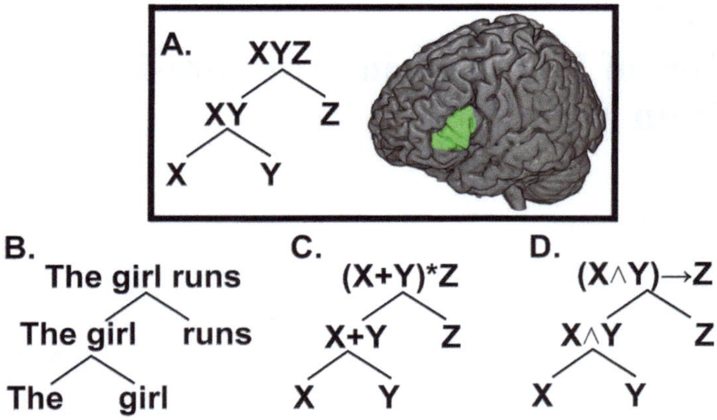

Fig. 5.1 The left inferior frontal gyrus (LIFG, highlighted in green) supramodal hierarchical parser hypothesis (partially adapted from Tettamanti and Weniger 2006)

it also characterizes several other aspects of human cognition such as logic reasoning, number and music cognition, action sequences and spatial relations, among others. As I will describe below, at least at an intuitive level, these seemingly distant domains of human cognition all appear to be organized at an abstract level and might therefore share, hidden behind a linear surface structure, the hierarchical and recursive features that are most commonly described by the syntactic trees built by linguists (see Fig. 5.1 for an example).

The objective of this chapter is to address the relationship (if any) between the mental computations that underlie the abstract structures we create when using natural language and those that underlie similar computations in other domains of human cognition. In what follows, I will first briefly trace the theoretical backdrop of this debate, and then present a dominant hypothesis concerning the role of language in human cognition, typically referred to as the supramodal hierarchical parser (SHP) hypothesis. According to this view a specific part of the human brain—traditionally considered to be a center for language processing—might in fact be involved in processing hierarchical structures across domains of human thought. In this chapter we review a number of functional neuroimaging experiments, specifically fMRI data, as they relate to the SHP hypothesis within the domains of logic reasoning and algebraic cognition. Finally the chapter will conclude by bringing together the different streams of evidence and evaluating the SHP hypothesis, as well as the overall debate concerning the role of language in human cognition.

5.2 Framing the Debate: Theoretical Background

Does language make us special? The extent to which the mechanisms of language contribute to shaping and organizing human cognition has been the focus of a longstanding debate. On the one hand, it is undeniable that language is one of the most characterizing aspects of the human mind. On the other hand, however, it is not clear whether the processes and properties of language are but one manifestation of the properties of our cognitive apparatus, or whether it is the emergence of language itself, in the human brain, that has endowed it with the ability to construct abstract representations, a computational infrastructure that might then have been relied upon by other domains of human cognition.

Taking a more general perspective, the debate concerning the intertwining of language and thought in the human mind has a long-standing tradition of proponents that fall somewhere in-between two extreme positions. According to one view, as formulated by Wilhelm von Humboldt, "language is the formative organ of thought. [...] Thought and language are therefore one and inseparable from each other" (Losonsky 1999, p. 99). As most frequently described, this view encompasses two complementary hypotheses. The first, often referred to as the *linguistic relativity* hypothesis, is a conjecture concerning the mechanism by which language exerts its influence on thought. As conceived by one of its most prominent proponents, Benjamin Lee Whorf, language shapes thought by providing the concepts around which perception of the world is organized:

> We dissect nature along lines laid down by our native languages. The categories and types that we isolate from the world of phenomena we do not find there because they stare every observer in the face; on the contrary, the world is presented in a kaleidoscopic flux of impressions which has to be organized by our minds and this means largely by the linguistic systems in our minds. We cut nature up, organize it into concepts, and ascribe significances as we do, largely because we are parties to an agreement to organize it in this way (Whorf 1940, p. 213).

Under this hypothesis, as noted by Edward Sapir, meanings are imposed upon us by "the tyrannical hold that linguistic form has on our orientation in the world" (Sapir 1931, p. 578), rather than discovered through experience. The second hypothesis, typically referred to as *linguistic determinism*, builds upon linguistic relativism and the observation of variability across languages, and concerns the effects of the influence of language on thought:

> The fact of the matter is that the 'real world' is to a large extent unconsciously built upon the language habits of the group. No two languages are ever sufficiently similar to be considered as representing the same social reality. The worlds in which different societies live are distinct worlds, not merely the same world with different labels attached. (Sapir 1929, p. 209)

At the other end of the spectrum sits a view according to which "thought is mediated by language-independent symbolic systems, often called the language(s) of thought. [...W]hen humans learn a language, they learn to express in it concepts already present in their prelinguistic system(s)" (Gelman and Gallistel 2004, p. 441). Under this hypothesis words are just symbols for mental experiences gained through experience with the world:

> Spoken words are the symbols of mental experience and written words are the symbols of spoken words. Just as all men have not the same writing, so all men have not the same speech sounds, but the mental experiences, which these directly symbolize, are the same for all, as also are those things of which our experiences are the images (Aristotle, *De Interpretatione* Chapter I, 16a).

This view of language is also evident in John Locke's work where words are conceived as signs of internal conceptions, and "stand as marks for the ideas within [the] mind, whereby they may be made known to others, and the thought of men's minds be conveyed from one to another" (Locke 1824, Book III, Chapter I). At its core, this view proposes that thought is prior to language (Pinker 1984). As explained by Li and Gleitman:

> Language has means for making reference to the objects, relations, properties, and events that populate our everyday world. It is possible to suppose that these linguistic categories and structures are more-or-less straightforward mappings from a preexisting conceptual space, programmed into our biological nature. [...] Humans invent words that label their concepts (Li and Gleitman 2002, p. 265).

Importantly, as explained by Gelman and Gallistel (2004), properties that are typically regarded as essential to language, such as compositionality, are in fact already present in our preexisting (i.e., prelinguistic) systems. The properties of language might thus just be one incarnation of the properties of thought (which might resemble those discussed by Fodor (1975)), no differently than the properties of other structure-dependent aspects of human cognition. In other words, we have the language we have in order to express the thoughts we have (Pinker and Jackendoff 2005).

5.3 The Relationship Between Language and Thought

5.3.1 The Algebraic Mind: Structure Dependence in Human Cognition

As mentioned in the introduction, one of the most fundamental hypothesis concerning how our minds process language is the idea that as we perceive serially ordered sequences of (linguistic) utterances, we spontaneously build non-linear (i.e., hierarchical) abstract representations which underlie our ability to assign meaning to a string of linguistic utterances. The psychological reality of these hierarchical constructs is well demonstrated by the first two lines of Lewis Carrol's famous poem Jabberwockie: *Twas brillig, and the slithy toves did gyre and gimble in the wabe*. Although semantically non-interpretable, the sentence feels structurally well-formed, while its reverse, for instance, despite featuring all the same words, does not: *Wabe the in gimble and gyre did toves slithy the and, brilling twas*. The relationship tying abstract structures and interpretation of linguistic statements is illustrated by Groucho Marx's famous statement *I once shot an elephant in my pajamas*. The hypothesis being, that the two possible interpretations of this sentence reflect two different abstract representations, each binding the elements of the sentence in a different way. If, for instance, our mental structure directly connects the verb phrase *I shot* to the propositional phrase *in my pajamas*, that would represent an understanding that the shoot*er* was wearing his night apparel as the events took place. On the other hand, if our mental representation of the sentence puts in direct relation the noun phrase *an elephant* with the propositional phrase *in my pajamas*, that would lead us to the (more puzzling) image that the shoot*ee* somehow managed to sneak into Groucho's nightwear before the fatal event.

Central to this view are a set of properties that lie at the heart of what we consider language (cf., Boeckx 2010, p. 32). The first property stresses the *abstract* nature of linguistic representations and relates to the so-called type-token distinction, which is to say the ability to recognize the difference between classes of elements, such as verbs and nouns, and the specific elements within a class, such as the verb *to go* or the noun *boy*. This property is crucial to the idea of an "algebraic mind" because it allows defining combinatorial rule mappings that apply over classes of items (i.e., types) rather than individual tokens, rendering the rules abstract. The second property, *compositionality* (or *structure dependence*) refers to the fact that the meaning of a complex expression is derived from the meanings of its constituents as well as the specific relationships by which they are bound together. This implies that there is a level of organization of the elements within a structure that confers meaning and exists independently of the semantics of the individual symbols, the psychological reality of which is well captured by Chomsky's famous example *Colorless green ideas sleep furiously*. It is this property of language that allows us to distinguish "between the boring news *dog bit man*, and the much more newsworthy *man bit dog*" (Boeckx 2010, p. 32). Third, *quantification* (or *bracketing*) refers to the ability to properly assign the right set of brackets around certain groups of elements within a statement and therefore understand the hierarchy by which elements within a structure bind. Finally, *recursion* refers to the property by which a rule can be applied to its own output or to the output of other rules, repeatedly, without any limit but for those imposed by the processing capacity of the individual (Corballis 1992).

While these properties are most prominently displayed in our use of natural language, they are also central to other domains of human cognition (cf., Tettamanti and Weniger 2006; Fadiga et al. 2009; Uddén and Bahlmann 2012), among others. As explained in Varley et al. (2005), for example, there exists an intuitive parallel between the interpretation of the two sentences "John kissed Jill" and "Jill kissed John" and the interpretation of the two algebraic statements "(5 – 3)" and "(3 – 5)." Despite the fact that, within each pair, the same elements are used, their different combination leads to different interpretation (e.g., who is kissing who, and whether the result is positive or negative, respectively). A similar analogy can be drawn between the recursive application of rules in language, as in the sentence "The man saw the boy who kicked the ball," and the recursive application of rules in algebra, as in "2 × (5 – 3)." In both examples structures can be embedded within structures of the same kind, and understanding of the hierarchical ordering by which elements within the structure bind is key to correct interpretation. Finally, both domains are infinitely generative in that it is possible for a finite set of elements to combine into a potentially unbound set of well-formed expressions. It is indeed always possible to generate a new sentence by prefixing, for example, "Mary thinks [that ...]" to any well-formed sentence just as it is always possible to generate a novel and meaningful algebraic statement by adding "2 + […]" to any well-formed algebraic expression. Whether this analogy is superficial or substantial is the topic of wide discussion, and will be the central concern of this chapter.

5.3.2 The "Supramodal Hierarchical Parser" Hypothesis

Given the *prima facie* similarities that can be drawn between language and other aspects of human thought, one might wonder whether a common set of computations lie at the heart of all structure-dependent cognition. This very intuition was already formulated in the work of Thomas Hobbes according to whom thinking amounted to performing arithmetic-like operations on internal structures (i.e., mental representations). While he did recognize that some forms of thought can exist outside of language, he believed that "mental processes where generality and orderly concatenation of thought are involved require the use of internal linguistic means" (Boeckx 2010). Linguistic computations might therefore be seen as central to our modes of thought, and as providing the very fabric of structure-dependent cognition. Following this view, it has been recently proposed that the human brain encapsulates, in the left inferior frontal gyrus (LIFG; and specifically its *pars opercularis* and *pars triangularis*), a supramodal hierarchical parser (SHP; Tettamanti and Weniger 2006; Fadiga et al. 2009; Uddén and Bahlmann 2012). The core of this proposal, which is depicted in Fig. 5.1, is the hypothesis that the LIFG is involved in the computations necessary for processing and representing abstract, hierarchical, "syntax-like," structured sequences across domains of human cognition. Following the explanation of Tettamanti and Weniger (2006), given three non-identical elements "X, Y, Z," their arrangement in a hierarchical fashion, as pictured in Fig. 5.1a, allows establishing univoque relations between the elements, thereby pinpointing one out of the three possible arrangement they can take, and thus assigning a specific meaning to the string. As shown in Fig. 5.1b, the same hierarchical structure can be employed to describe the natural language statement *The girl runs*. Specifically, the first two elements, the determiner *the* and the noun *girl*, are first bound together (into a noun phrase) and then, as a unit, bound to the verb *runs*, thereby imparting a specific interpretation to the sentence. Importantly, any other ordering of the three elements, obtained by permuting their position, would either yield an ill-formed sentence, or a well-formed sentence with a different intension. This same hierarchical structure can also be employed to describe, as shown in Fig. 5.1c, the algebraic statement "$(X + Y) \times Z$" where the first elements are bound together by an addition operator, and the result of that operation is then bound to the third element by a multiplication operator. As for the linguistic statement, if the hierarchical structure binding the elements were different, the statement would have a different interpretation (i.e., result). A similar case can be made for the logic statement "$(X \wedge Y) \rightarrow Z$" (which translates to the natural language statement "If X and Y then Z"), as depicted in Fig. 5.1d, as well as several other domains of human thought including music cognition, spatial relations, and action sequences (e.g., Tettamanti and Weniger 2006; Fadiga et al. 2009).

Although the specific function of the LIFG is still a matter of debate (cf., Hagoort 2005; Grodzinsky and Santi 2008), several lines of evidence suggest that it is crucial to the processing of natural language (cf., Bookheimer 2002). Particularly relevant to our discussion, this region of the brain is consistently found to be active for complex syntactic statements, such as object-embedded sentences (e.g., "The man that the girl is talking to is happy"), as compared to simpler subject-embedded sentences (e.g., "The girl that is talking to the man is happy"; cf., Just et al. 1996). Indeed, understanding a sentence's argument hierarchy construction (i.e., who did what to whom; Bornkessel et al. 2005) as well as whether thematic roles remain unchanged through syntactic transformations (e.g., active to passive; Monti et al. 2009, 2012) consistently recruits the LIFG (although not exclusively). Furthermore, the LIFG appears to be recruited for processing the long-distance dependencies and hierarchical structures (Friederici 2004) that are at the heart of natural language (Lees and Chomsky 1957). Although it is clear that the LIFG plays a crucial role in natural language (and its "structural" aspect in particular), there are several contrasting hypotheses concerning which computation(s) are specifically embedded within the neural circuitry of this region. While according to some this region might carry out computations specific to establishing long-range dependencies (Friederici 2004) and syntactic movement (Grodzinsky and Santi 2008), it has also been suggested that it might be involved in unifying lexical information (Hagoort 2005) or, more generally, selecting among competing representations (Novick et al. 2010). I should stress, however, that the question of which role (if any) linguistic computations play in other domains of cognition is neutral with respect to this debate. For, inasmuch as it is agreed that whichever the specific computation carried out within the LIFG this is core to our processing of language, all that matters is evaluating whether this neural circuitry is also involved in the computations of other aspects of human cognition.

Of course, there is certain a definitional component to establishing whether a given domain of human thought is "linguistic." For, how language is defined in the human brain might significantly affect which aspects of human cognition might be considered as resting on linguistic computations. As discussed in other chapters of this book, language (broadly conceived) encompasses a rich and wide set of cognitive processes. From a neural point of view, linguistic stimuli can thus elicit activations in several areas outside the LIFG as well as other "traditional" perisylvian language regions. Several fMRI and clinical studies, for example, have highlighted the role of motor cortices in processing action-related words (Hauk et al. 2004), medial temporal lobe regions in semantic processing (Hoenig and Scheef 2005), right hemispheric fronto-temporal areas in processing prosodic cues (Wildgruber et al. 2006), temporo-parietal and subcortical reward-related regions in processing humor (Bekinschtein et al. 2011), among many others. In fact, if the pragmatics of message selection (Grice 1991) is counted as a core linguistic capacity, then virtually any neural area implicated in cognition could be considered a language structure (Monti et al. 2009). In the following discussion, however, I will focus on the set of processes underlying the construction of rule-governed relationships that allow generating the unbound range of possible expressions within a language (Chomsky 1983).

5.4 The Role of Language in Structure-Dependent Cognition

5.4.1 Disentangling "Language" and "Thought" with fMRI

Before discussing the neuroimaging evidence concerning the role of language in structure-dependent cognition, it is worth reviewing some of the crucial features of functional magnetic resonance imaging (fMRI) as it is employed today to uncover the neural basis of human cognition. In particular, it is important to note that the fMRI signal (typically referred to as the blood oxygenation level dependent signal, BOLD; Ogawa et al. 1990) is difficult to interpret per se. Knowing that a mental activity elicits a BOLD signal of, say, 850 units in a given part of the brain is not very meaningful. More meaningful is the comparison of the BOLD signal between two different tasks. Hence, most task-based fMRI studies are based on the so-called subtraction principle whereby the metabolic response to a task of interest is compared to the metabolic response to a control (or "baseline") task. If this latter task contains all the same cognitive processes as the task of interest, except for the one mental process of interest (often referred to as the "pure insertion" hypothesis), subtracting the metabolic response to the baseline task from that of the main task should isolate the metabolic response specific to the cognitive process of interest. Evaluating an experiment's baseline is thus critical to correctly interpreting a functional neuroimaging result. Imagine, for example, being interested in the neural basis of single word repetition. As a main (or target) task, one might present visually a set of words, one at a time, and ask participants to repeat them. As a baseline task, one might decide to employ periods of rest during which the participant is not performing any (overt) task. Subtracting the metabolic activity observed during the latter periods from that observed during the main task will likely isolate a wide set of neural foci including both the neural substrate of single words repetition and several other ancillary processes tied to processing visual stimuli as well as words. This baseline task is highly sensitive, because it captures all the neural structures that are elicited by the target task, but not very specific, because it captures many processes that are not specifically related to the cognitive process of interest. In other words, of all the regions of the brain that might be uncovered by this subtraction, it is difficult to tell which are directly involved in word repetition and which are tied to the many other processes that factor into the target task. Given the same main task, adopting a baseline task in which participants are presented with strings of letters that do not form a meaningful word might be more effective in filtering out basic visual processes from the activations elicited by the main task. Nonetheless, the subtraction is still likely to uncover phonetic and semantic processes tied to reading meaningful words as well as the process of interest (i.e., single-word repetition). One might thus employ a baseline task in which subjects are presented with single words, as in the target task, and asked to read them. This baseline should be very effective in filtering out from the main task all the ancillary activations related to processing visual stimuli as well as the phonetics and semantics of reading single words. Ideally, the comparison would thus pinpoint only regions involved in word repetition. However, what if participants spontaneously and automatically repeat the words subvocally as they read them? In this case, the baseline could elicit the same neural substrate as the target task filtering out, partially or entirely, the metabolic response related to the cognitive process of interest.

While the above examples might appear extreme and unlikely to exist in actual practice, as I will discuss in the next section, similar circumstances continuously arise in cognitive neuroscience research often resulting in substantial divergence of results across studies. Furthermore, this issue is particularly severe within the domain of higher cognitive functions where eliciting the process(es) of interest (e.g., reasoning) often requires eliciting several other ancillary processes.

5.4.2 Language and Logic Reasoning

Background Deductive reasoning is the attempt to draw secure conclusions from prior beliefs, observations and suppositions (Monti and Osherson 2012). This aspect of human cognition has been the focus of vigorous investigation within the fields of philosophy and psychology (e.g., Beall and van Fraassen 2003; Osherson and Falmagne 1975). It is typically regarded as a central feature of human intelligence (Rips 1994), although some forms of deduction (e.g., transitive inference) have also been reported in other species (e.g., Grosenick et al. 2007). With respect to the role of language in deductive reasoning, different views have been expressed.[1] On some accounts, language plays a central role in the deductive inference making process (Polk and Newell 1995). According to others, reasoning is fundamentally based on processes other than the syntactic interpretation of sentences (Cheng and Holyoak 1985; Osherson and Falmagne 1975).[2] In considering the neurobiology of deductive competence (as well as algebraic cognition—see next section), it is important to distinguish two potential roles for linguistic processing. At a minimum, the (verbal) stimuli typically employed to elicit deductive reasoning must be apprehended

[1] As described below, given that deductive reasoning is most often elicited by the means of verbal stimuli, it is trivial that linguistic processes are needed to apprehend the stimuli. What is under discussion here is whether linguistic processes play a role in deductive reasoning beyond the initial encoding of verbal materials.

[2] It might be worth clarifying that so-called Mental Rules theories of deduction (e.g., Osherson and Falmagne 1975), despite being sometimes portrayed as language based (see Goel et al. 1998, 2000), might in fact be better understood as describing deductive inference as a "syntax-like," algebraic, computation, rather than a linguistic one (cf., Monti et al. 2007).

before deduction can take place. At a neural level, language and reading areas (Price 2000; Bookheimer 2002) would thus be expected to be involved in this stage. What is more controversial (and under discussion in this chapter) is whether language plays a part in the subsequent inferential process itself.

Neuroimaging Studies of Deductive Reasoning Overall, neuroimaging studies of reasoning have defended a variety of positions including the thesis that all deductive reasoning is left-hemispheric and language based (e.g., Goel et al. 1997, 1998; Reverberi et al. 2007), along with the contrary suggestion that none of it is (e.g., Goel and Dolan 2001; Parsons and Osherson 2001; Knauff et al. 2003). Yet, other studies have been interpreted as supporting a "dual process" view of deduction according to which, depending on whether the reasoner has prior beliefs over, or familiarity with, the contents of the argument she is reasoning about, language resources may or may not be recruited (e.g., Goel and Dolan 2003). This dramatic variance of results highlights the complexities of disentangling "thought" from linguistic processes using correlational methods such as fMRI, and is, to a significant extent, tied to the subtraction problem discussed in the previous section. Knauff et al. (2003), for example, recorded the metabolic response of healthy volunteers while they judged whether each of a number of arguments featuring two premises and one conclusion, were deductively valid (for example: "The dog is cleaner than the cat." "The ape is dirtier than the cat." Does it follow: "The dog is cleaner than the ape?"). Comparison of the metabolic response during the target task to that observed during rest periods uncovered activations in some left hemispheric language regions, among others. As discussed above, due to the non-specific nature of the baseline task, it difficult to assess whether the involvement of posterior perisylvian language regions reflects the engagement of linguistic resources during the deductive inference process or during the initial processing of verbal stimuli. In a set of pioneering neuroimaging studies, Goel et al. (1997, 1998) employed a baseline task in which subjects were asked to determine how many of the three sentences in a given argument had people as their subject. While this baseline does, to some extent, filter out ancillary processes related to encoding visual and verbal stimuli, the minimal amount of linguistic processing required is likely to be less than that required to read the same sentences in view of inferential reasoning. Thus, again, it is difficult to tell whether the observed activations in linguistic regions reflect simple reading or the involvement of linguistic mechanisms in deductive reasoning. Other experimental design factors, such as the timing of the task of interest (as well as the baseline one), can also substantially effect the interpretation of neuroimaging findings. Goel et al. (2000), for example, employ a baseline task that is isomorphic to the target task but included a conclusion unrelated to the premises. To illustrate, consider the two deductive trials presented in Table 5.1 (each consisting of two premises and one conclusion).

The idea of comparing the metabolic activity in response to the two arguments is very clever because the status of a trial (with respect to being a target or baseline trial) depends on whether the conclusion is related to the premises, as in Argument #1, allowing deduction to take place, or not, as in Argument #2. The participant,

Table 5.1 Sample deductive stimuli from Goel et al. (2000). ("$P_{1,2}$" indicate Premise 1 and Premise 2; "C" indicates the Conclusion of the argument.)

	Argument #1	Argument #2
P_1	All poodles are pets	All poodles are pets
P_2	All pets have names	All pets have names
C	[Therefore] All poodles have names	[Therefore] No napkins are white

however, is unaware of such distinction and performs all trials under the same set of instructions, namely to assess whether the conclusion follows from the premises. This experimental setup, however, has two very problematic and unwanted consequences. First, the presence of extraneous materials early in the conclusion statement (i.e., napkins) is sufficient for the participant to recognize, with very little reading, the invalidity of the trial (and therefore that it is a baseline trial). Thus, as for the two previous experiments discussed above, the baseline task might not sufficiently filter out linguistic processes tied to sentence reading during deductive trials. Second, the slow sequential presentation of each statement, serially added to the display at 3 s intervals, allows deductive processes to start taking place as soon as the second premise is displayed and, crucially, before the conclusion is presented. Thus, until the conclusion is presented, target and baseline trials might elicit comparable amounts of deductive reasoning. As a result, this baseline task may subtract essential elements of deductive reasoning from deduction trials, while not adequately filtering reading activations (cf., Monti and Osherson 2012). It is not surprising, then, that the authors report engagement of linguistic regions in the LIFG for the target task (compared to the baseline).

A Language-Independent Network for Deductive Reasoning As the above discussion illustrates, characterizing the neural substrate of deductive reasoning presents several complexities which have prevented the field from reaching a consensus on what role (if any) language plays in this aspect of human thought. In a recent series of experiments, however, the case has been made for deductive reasoning recruiting a language-independent distributed network of brain regions (see Monti and Osherson 2012, for a review). In an attempt to avoid many of the experimental pitfalls described above, Monti et al. (2007) adopted a "cognitive load" design in which participants were instructed to assess whether each of a number of logic arguments were deductively valid. Half the arguments were simple to assess (e.g., "If the block is either round or large then it is not blue." "The block is round." "The block is not blue."), whereas the other half were more complex (e.g., "If the block is either red or square then it is not large." "The block is large." "The block is not red."). From a cognitive perspective, complex and simple deductions can be expected to recruit the same kind of mental operations, but in different number, repetition, or intensity. If linguistic structures are involved in the inferential process, complex deductions should recruit them significantly more than simple ones. On the other hand, if the role of language is confined to initial encoding of stimuli, simple inferences can be expected to require similar levels of reading compared to their complex counterparts. This expectation is reinforced by the fact that the statements included in the simple and complex arguments are matched for linguistic complexity (as the two sample inferences above demonstrate) and can be expected to prompt for similar amounts of initial language processing. Thus, should any language-related activation be apparent, it cannot be considered to reflect differences in initial reading or comprehension. Subtraction of the metabolic response observed during simple trials from that observed during complex trials was thus expected to adequately filter-out the initial reading of verbally presented materials while revealing areas of the brain correlating with increased deductive reasoning. The authors reported two main findings. First, the complex minus simple subtraction did not reveal any activation in the LIFG supramodal hierarchical parser (as well as in posterior temporal regions), indicating that although the region was active at the beginning of each trial, in correspondence with initial reading, it was equally recruited by simple and complex inferences. Second, the subtraction uncovered a distributed network of areas spanning regions that are believed to perform cognitive operations that sit at the "core" of deductive reasoning (in frontopolar and fronto-medial cortices) as well as several other "cognitive support" frontal and parietal regions known to be related to working memory and spatial attention process. Using a different approach, Monti et al. (2009) compared logic inferences based on sentential connectives (i.e., "if … then …," "and," "or," "not") to inferences based on the syntax and semantics of ditransitive verbs (e.g., "give," "say," "take"). To illustrate, a valid linguistic inference might include the premise "X gave Y to Z" and the conclusion "Z was given Y by X." Similarly, a valid logic inference might feature the premise "If both X and Y then not Z" and the conclusion "If Z then either not X or not Y." In this design, logic and linguistic arguments were each compared to a matched baseline in which the very same sentences evaluated for inferential validity were also evaluated for grammatical well-formedness. Occasionally, in order to ensure that participants fully encoded the sentences during the baseline trials, they were presented with "catch trials" in which statements contained grammatical anomalies (e.g., "X gave to Y to Z"). If logical inference is based on mechanisms of natural language that go beyond mere reading for meaning, then the comparison of each type of inference to its matched baseline should uncover common activations in regions known to underlie linguistic processing, and particularly within the LIFG SHP. Subtraction of the metabolic response during grammatical judgments from linguistic inferences uncovered extensive activation within the LIFG, confirming its role in evaluating semantic equivalence of distinct sentences (Dapretto and Bookheimer 1999), morphological processing (Sahin et al. 2006), detecting semantic roles (Bornkessel et al. 2005), transforming sentence syntax (Ben-Shachar et al. 2003). Conversely, when the same comparison was performed on logic arguments, no activity was detected in the LIFG (or posterior temporal perisylvian language regions), indicating that logic reasoning does not recruit language resources beyond what is necessary for simple reading. When the logic and linguistic inference tasks were compared directly (over the full brain, as well as on a region-by-region basis), only the latter were shown to engage the LIFG (as well as posterior temporal perisylvian regions). Furthermore, the activations detected during the logic inference trials replicated exactly those seen, with a different task, in Monti et al. (2007) (see Monti and Osherson 2012, Fig. 5.1).

Overall, several studies have failed to uncover activation within the LIFG during deductive reasoning tasks (e.g., Noveck et al. 2004; Canessa et al. 2005; Fangmeier et al. 2006; Prado and Noveck 2007; Rodriguez-Moreno and Hirsch 2009; Prado et al. 2010b). These findings can be interpreted as implying that the role of language is confined to initial encoding of verbal statements into mental representations suitable for the inferential calculus. The representations themselves, as well as the deductive operations, however, are not linguistic in nature.

Neuropsychological Evidence Despite the above data, the view just formulated is still very debated inasmuch as a number of fMRI studies have reached the opposite conclusion (e.g., Reverberi et al. 2007, 2010; Prado et al. 2010a). As discussed in Monti and Osherson (2012), several factors relating to experimental design can explain this fracture within the literature. To address this point, however, it might be worth considering evidence from the neuropsychological literature, which has the distinct advantage, over neuroimaging methods, to uncover causal relationships between cognitive processes and neural circuitry. In particular, Reverberi et al. (2009) assessed inferential abilities in patients with left lateral frontal damage, right lateral frontal damage, or medial frontal damage. Interestingly, patients with damage in right frontal cortex exhibited no apparent difficulty in assessing the validity and judging the difficulty of deductive inferences. Patients with damage in left frontal cortex, instead, appeared impaired in assessing deductive problems, but only inasmuch as their working memory was affected (i.e., patients with intact working memory, as tested with standard neuropsychological measures, were able to correctly perform the inferential task). Finally, patients with medial frontal damage were unable to solve deductive inferences (a finding recently replicated by Koscik and Tranel (2012)). Two aspects of these results are crucial. First, no patient had damage to the LIFG, nor appeared to have language deficits, a fact difficult to reconcile with the claim that language processes are at the heart of this domain of cognition. Second, the cortical damage that impaired deductive reasoning falls in the same prefrontal areas that have been previously characterized as "core" to deductive inference (Monti et al. 2007, 2009; Rodriguez-Moreno and Hirsch 2009). Overall, then, although the question is still very debated, neuropsychological evidence suggests that the neural mechanisms within the LIFG are not sufficient for supporting this kind of structure-dependent cognition, while the fMRI evidence, together with the patient data, suggest that medial and polar frontal cortex might be necessary and sufficient (perhaps among other regions of the brain) for processing the hierarchical dependencies imposed by logic structure.

5.4.3 *Language and Arithmetic Cognition*

Background The relationship between natural language and mental arithmetic has also long been debated (e.g., Spelke and Tsivkin 2001; Gelman and Gallistel 2004). In particular, as discussed above (and more extensively in Varley et al. (2005)), there is a certain analogy between the operations of natural language and those of mental arithmetic. Indeed, it has been argued that both language and number rely on a recursive computation that exploits the same neural mechanism operating over linguistic structures (Hauser et al. 2002). Recursion might thus have evolved over time from a process that was highly domain specific (to natural language) to a domain general process that gave the human mind the unique ability to use recursion to solve non-linguistic problems such as numerical manipulation. A similar view is also embraced by Spelke and Tsivkin (2001) who stated that natural language is the "most striking combinatorial system" of the human mind and that formal mathematics might be one of this system's "richest and most dramatic outcomes" (p. 84). The view that arithmetic reasoning might have co-opted the recursive machinery of language is also explicit in Chomsky (1998) where he argues that the human faculty for arithmetical reasoning can be thought of as being abstracted from language and that it operates by "preserving the mechanisms of discrete infinity and eliminating the other special features of language" (p. 169). Similarly, Fitch et al. (2005) state that the only clear demonstrations that recursion operates in human cognitive domains come from mathematical formulas and computer programming, which clearly employ the same reasoning processes that language does. Overall, as for other domains of human cognition, the debate is whether "the generative power of grammar might provide a general cognitive template and a specific constitutive mechanism for 'syntactic' mathematical operations involving recursiveness and structure dependency" (Varley et al. 2005, p. 3519).

Neuroimaging Studies of Number Cognition In a landmark study by Dehaene et al. (1999), the relationship tying language and mathematical knowledge was addressed with a joint behavioral and neuroimaging approach. In the behavioral study, bilingual speakers learned to do arithmetics, including exact and approximate calculations, in one of two languages. When tested on trained and novel exact calculations, participants exhibited a language-switching penalty, which is to say, when the language in which calculations were trained mismatched the language in which they were later tested, participants exhibited slower reaction times. Under the same circumstances, however, approximate calculations did not exhibit a comparable language switching cost, suggesting that only exact arithmetic knowledge hinges on linguistic knowledge/mechanisms. When the two tasks were compared using fMRI, approximate calculations recruited mainly parietal regions, whereas exact calculations recruited regions associated with some aspects of linguistic processing, in the angular gyrus as well as the inferior frontal regions (although in an area that appears more frontal than the LIFG SHP). In a subsequent study, Stanescu-Cosson et al. (2000) compared neural activation for exact and approximate calculation using both small and large numbers. Approximate calculations elicited activity in the bilateral intraparietal sulci (among several others), confirming the view that linguistic resources are not recruited by this aspect of number cognition. Exact calculation, on the other hand, mainly recruited (among others) the angular gyrus as well as the anterior section of the inferior frontal gyrus. However, as for the results presented by Dehaene et al. (1999), the region of the inferior frontal gyrus that was found active is clearly anterior to that observed during syntactic processing of natural language and may thus be related to control of verbal retrieval processes implemented in more posterior cortico-subcortial verbal networks (as explained in Stanescu-Cosson et al. 2000, p. 2252). The authors do point out that the LIFG proper was found to be specifically recruited by exact calculations with large numbers; however, they speculate that its involvement is mainly a function of the increased effortfulness associated with lesser known facts involved in operations with large numbers.

Overall, the results presented above, together with several others, have coalesced in a coherent view according to which language may well play a role in exact calculation, but in connection with the verbal coding of rote exact arithmetic facts (Dehaene et al. 2003, 2004).

A Language-Independent Network for Processing Algebraic Structure Although most studies in this domain are primarily concerned with the neural basis of the representation of quantity, numbers, and calculation, two concurrent studies investigated the role of language in processing and manipulating the syntax-like structures of algebra (Maruyama et al. 2012; Monti et al. 2012, respectively). Maruyama et al. (2012), for example, looked at the cortical representation of simple algebraic expressions such as "1 + (4 (2 + 3))" to assess whether the neural structures responsible for parsing these nested structures are indeed shared with the neural mechanisms responsible for computing the syntax of linguistic expressions. In this experiment, participants were shown strings of algebraic symbols variously arranged so as to form expressions with high levels of nesting (e.g., "4 + (1 (3 + 2))") versus strings with little or no nesting (e.g., ") (3 2) + 4(+1") or algebraically meaningless strings (e.g., "4 +)3)(+2(1"). Importantly, participants were not asked to resolve the equations, but rather just to encode the expressions sufficiently for a probe matching task (i.e., to decide whether the string matches a probe presented at a short delay). Contrary to the LIFG supramodal parser hypothesis, all the regions that were increasingly recruited by greater nesting fell outside the traditional left perisylvian language regions, and mainly included occipital, temporal, and inferior parietal regions, indicating that processing syntactically complex algebraic expressions does not rely on the LIFG supramodal parser or other traditional language regions. In that same year, Monti et al. (2012) addressed the question of whether manipulating the syntax-like structure of algebraic expressions relies on the same neural structures required to manipulate the syntax of natural language statements. In a design similar to that used in Monti et al. (2009), discussed above, participants were presented with pairs of natural language statements (e.g., "X gave Y to Z" and "Z was given Y by X") and algebraic statements (e.g., "X minus Y is greater than Z" and "Z plus Y is smaller than X"). Participants were instructed to evaluate whether the statements within each pair were equivalent. Although the two tasks are psychologically very similar, judging equivalence in the former kind of pairs depends on whether the principal verb assigns the same semantic roles (i.e., who did what to whom) to X, Y, and Z across a syntactic transformation. Conversely, judging the equivalence in the algebraic pairs depends on the properties of elementary algebraic operations (i.e., addition, subtraction) and relations (i.e., equality, inequality). Therefore, if left IFG truly acts as a supramodal parser of hierarchical structure, this region should be equally involved in processing and manipulating the hierarchical dependencies of linguistic and algebraic expressions. As a baseline, participants were shown the same statements but asked to assess whether they were grammatically correct (as in Monti et al. 2009). When the linguistic trials were analyzed, subtraction of the grammar judgment task from the equivalence task revealed extensive activation in the LIFG as well as other perisylvian linguistic areas, as expected. However, when the same comparison was performed on the algebraic pairs, no activation was detected in the supramodal parser (or any other perisylvian language region), indicating that beyond initial reading and comprehension of stimuli, the neural substrate of language does not intervene in algebraic reasoning (consistent with the findings of Monti et al. (2009), Maruyama et al. (2012)). Conversely, extensive activation was detected in the infra-parietal sulci, consistent with the number cognition literature (cf., Dehaene et al. 2003).

Neuropsychological Evidence In the domain of mental arithmetic there is relatively rich neuropsychological and developmental disorder evidence that supports the neuroimaging findings. Indeed, some patients with semantic dementia and global aphasia have been shown to retain mathematical competence (e.g., Cappelletti et al. 2001; Delazer et al. 1999, respectively), while individuals with, for example, developmental discalculia and William's Syndrome have been shown to have severe impairment in the number domain while retaining normal language skills (Butterworth 2008; Bellugi et al. 1993, respectively). One case in particular demonstrated that the ability to process the structured hierarchy of algebraic expression can be retained in patients suffering from agrammatic aphasia, which is to say patients who appear to be unable to process the structured hierarchy of natural language (Varley et al. 2005). Indeed, when three patients with extensive left hemispheric damage were presented with the reversible sentences such as "The lion killed the man" they were unable to match it to the appropriate figure (when having to choose between the figure of a man killing a lion and that of a lion killing a man). However, when presented with reversible algebraic expressions, such as "$(3 - 5)$", patients had no trouble judging whether the result was positive or negative. This dissociation indicates that while the patients appeared to have lost structure sensitivity in the domain of language, they did retain it in the domain of algebra. Similarly, patients also appeared to have lost the ability to perform bracketing in language, as assessed by their inability to judge whether a sentence is grammatical or not. Nonetheless, they did retain the ability to correctly resolve algebraic expressions observing the hierarchy expressed by parenthetical structures. Overall, this brief overview of some of the cardinal results in the neuropsychological literature suggest, in accord with the neuroimaging evidence, that the structured hierarchy of algebra is not processed, in the human brain, by the neural mechanisms of natural language.

5.5 Discussion

Overall, at least with respect to the domains of deductive reasoning and mental algebra, the neuroimaging evidence fails to support the "supramodal hierarchical parser" hypothesis, and thus the view that the LIFG encapsulates neural circuitry tuned to detect and represent complex hierarchical dependencies regardless of the specific domain of cognition (Tettamanti and Weniger 2006; Fadiga et al. 2009). Furthermore, the above evidence also suggests that other left hemispheric perisylvian regions, in posterior temporal cortices, typically considered to be at the heart of language processing, also do not contribute to the processing and manipulation of deductive and algebraic hierarchical structures. The neuropsychologic and developmental disorder literature confirms the view that language mechanisms are not sufficient for deductive reasoning and number cognition. Indeed, as discussed above, the patients with fronto-medial damage described by Reverberi et al. (2009) and Koscik and Tranel (2012), as well the dyscalculic patients discussed in Butterworth (2008), are unable to process the structured hierarchies of deductive inference and number cognition despite showing normal language skills. Furthermore, the patients described by Varley et al. (2005), who retain the ability to understand the structured-hierarchy of algebra while being at chance in language comprehension, suggest that language is also not necessary to access the representations and computations of algebra. (To date, there is no parallel finding in the domain of deductive inference, although Varley and Siegal (2000), report the case of an agrammatic aphasic patient who, despite profound language impairment, could successfully perform causal reasoning.) Taken together, the presented data suggest that the involvement of language in structure-dependent cognition might be limited to what Polk and Newell (1995) termed a "transduction" role. That is, the mechanisms of language might be involved in decoding verbally presented information into representations suitable for algebraic and deductive computations (and encoding their output into language, if needed). The representations and "syntax-like" operations themselves, however, are not in linguistic format.

While the above data establishes that structure-dependent cognition is not parasitic on language *in the mature cognitive architecture*, it is still possible that the generative power of language plays an enabling role through ontogeny (e.g., Bloom 1994; Spelke 2003), or has played such a role through phylogenetic history (e.g., Corballis 1992). The case for homology, however, faces the complication of having to explain how the neural mechanisms of language extended into other domains, re-implemented their circuitry in distant brain regions (fronto-polar and fronto-medial cortices for deductive inference, Monti et al. 2009; Rodriguez-Moreno and Hirsch 2009; and the infra-parietal sulci for algebra, Monti et al. 2012; Maruyama et al. 2012), to then entirely disconnect from them.

In conclusion, while there is a certain beauty and efficiency in the hypothesis that language, the most characterizing and striking aspect of the human mind, provided the computations enabling generative cognition across domains of cognition, the available data does not support this view. In the words of Albert Einstein, "[w]ords and language do not seem to play any part in my thought processes. The physical entities which seem to serve as elements in thought are signs and images which can be reproduced and combined at will" (Hadamard 1954).

Acknowledgments This work was funded by the James S. McDonnell Foundation Scholar Award. Parts of this work have appeared in Monti et al. (2009, 2012), and Monti and Osherson (2012).

References

Beall, J. C., & van Fraassen, B. C. (2003). *Possibilities and paradox: An introduction to modal and many-valued logic*. Oxford: Oxford University Press.

Bekinschtein, T. A., Davis, M. H., Rodd, J. M., & Owen, A. M. (2011). Why clowns taste funny: The relationship between humor and semantic ambiguity. *Journal of Neuroscience, 31*, 9665–9671.

Bellugi, U., Marks, A. B., Bihrle, A., & Sabo, H. (1993). Dissociation between language and cognitive functions in Williams syndrome. In D. Bishop & K. Mogford (Eds.), *Language development in exceptional circumstances* (p. 177–189). Hove: Psychology.

Ben-Shachar, M., Hendler, T., Kahn, I., Ben-Bashat, D., & Grodzinsky, Y. (2003). The neural reality of syntactic transformations evidence from functional magnetic resonance imaging. *Psychological Science, 14*, 433–440.

Bloom, P. (1994). Generativity within language and other cognitive domains. *Cognition, 51*, 177–189.

Boeckx, C. (2010). *Language in cognition: Uncovering mental structures and the rules behind them*. Malden: Wiley.

Bookheimer, S. (2002). Functional MRI of language: New approaches to understanding the cortical organization of semantic processing. *Annual Review of Neuroscience, 25*, 151–188.

Bornkessel, I., Zysset, S., Friederici, A. D., von Cramon, D. Y., & Schlesewsky, M. (2005). Who did what to whom? The neural basis of argument hierarchies during language comprehension. *Neuroimage, 26*, 221–233.

Butterworth, B. (2008). Developmental dyscalculia. In J. Reed & J. Warner-Rogers (Eds.) *Child neuropsychology: Concepts, theory, and practice* (pp. 357–374). Chichester: Wiley.

Canessa, N., Gorini, A., Cappa, S. F., Piattelli-Palmarini, M., Danna, M., Fazio, F., et al. (2005). The effect of social content on deductive reasoning: An fMRI study. *Human Brain Mapping, 26*, 30–43.

Cappelletti, M., Butterworth, B., & Kopelman, M. (2001). Spared numerical abilities in a case of semantic dementia. *Neuropsychologia, 39*, 1224–1239.

Cheng, P. W., & Holyoak, K. J. (1985). Pragmatic reasoning schemas. *Cognitive Psychology, 17*, 391–416.

Chomsky, N. (1983). Noam Chomsky's views on the psychology of language and thought. In R. W. Rieber (Ed.), *Dialogues on the psychology of language and thought: Conversations with Noam Chomsky, Charles Osgood, Jean Piaget, Ulric Neisser, and Marcel Kinsbourne*. New York: Plenium.

Chomsky, N. (1998). *Language and problems of knowledge: The Managua lectures* (Vol. 16). Cambridge, MA: MIT Press.

Corballis, M. C. (1992). On the evolution of language and generativity. *Cognition, 44*, 197–226.

Dapretto, M., & Bookheimer, S. Y. (1999). Form and content: Dissociating syntax and semantics in sentence comprehension. *Neuron, 24*, 427–432.

Dehaene, S., Molko, N., Cohen, L., & Wilson, A. J. (2004). Arithmetic and the brain. *Current Opinion in Neurobiology, 14*, 218–224.

Dehaene, S., Piazza, M., Pinel, P., & Cohen, L. (2003). Three parietal circuits for number processing. *Cognitive Neuropsychology, 20*, 487–506.

Dehaene, S., Spelke, E., Pinel, P., Stanescu, R., & Tsivkin, S. (1999). Sources of mathematical thinking: Behavioral and brain-imaging evidence. *Science, 284*, 970–974.

Delazer, M., Girelli, L., Semenza, C., & Denes, G. (1999). Numerical skills and aphasia. *Journal of the International Neuropsychological Society, 5*, 213–221.

Fadiga, L., Craighero, L., & D'Ausilio, A. (2009). Broca's area in language, action, and music. *Annals of the New York Academy of Sciences, 1169*, 448–458.

Fangmeier, T., Knauff, M., Ruff, C. C., & Sloutsky, V. (2006). fMRI evidence for a three-stage model of deductive reasoning. *Journal of Cognitive Neuroscience, 18*, 320–334.

Fitch, W. T., Hauser, M. D., & Chomsky, N. (2005). The evolution of the language faculty: Clarifications and implications. *Cognition, 97*, 179–210.

Fodor, J. A. (1975). *The language of thought* (Vol. 5). Cambridge, MA: Harvard University Press.

Friederici, A. D. (2004). Processing local transitions versus long-distance syntactic hierarchies. *Trends in Cognitive Sciences, 8*, 245–247.

Gelman, R., & Gallistel, C. R. (2004). Language and the origin of numerical concepts. *Science, 306*, 441–443.

Goel, V., Buchel, C., Frith, C., & Dolan, R. J. (2000). Dissociation of mechanisms underlying syllogistic reasoning. *Neuroimage, 12*, 504–514.

Goel, V., & Dolan, R. J. (2001). Functional neuroanatomy of three-term relational reasoning. *Neuropsychologia, 39*, 901–909.

Goel, V., & Dolan, R. J. (2003). Explaining modulation of reasoning by belief. *Cognition, 87*, B11–B22.

Goel, V., Gold, B., Kapur, S., & Houle, S. (1997). The seats of reason? An imaging study of deductive and inductive reasoning. *NeuroReport, 8*, 1305–1310.

Goel, V., Gold, B., Kapur, S., & Houle, S. (1998). Neuroanatomical correlates of human reasoning. *Journal of Cognitive Neuroscience, 10*, 293–302.

Grice, H. P. (1991). *Studies in the Way of Words*. Cambridge, MA: Harvard University Press.

Grodzinsky, Y., & Santi, A. (2008). The battle for Broca's region. *Trends in Cognitive Sciences, 12*, 474–480.

Grosenick, L., Clement, T. S., & Fernald, R. D. (2007). Fish can infer social rank by observation alone. *Nature, 445*, 429–432.

Hadamard, J. (1954). *An essay on the psychology of invention in the mathematical field*. Mineola: Courier Corporation.

Hagoort, P. (2005). On Broca, brain, and binding: A new framework. *Trends in Cognitive Sciences, 9*, 416–423.

Hauk, O., Johnsrude, I., & Pulvermüller, F. (2004). Somatotopic representation of action words in human motor and premotor cortex. *Neuron, 41*, 301–307.

Hauser, M. D., Chomsky, N., & Fitch, W. T. (2002). The faculty of language: What is it, who has it, and how did it evolve? *Science, 298*, 1569–1579.

Hoenig, K., & Scheef, L. (2005). Mediotemporal contributions to semantic processing: fMRI evidence from ambiguity processing during semantic context verification. *Hippocampus, 15*, 597–609.

Just, M. A., Carpenter, P. A., Keller, T. A., Eddy, W. F., & Thulborn, K. R. (1996). Brain activation modulated by sentence comprehension. *Science, 274*, 114.

Knauff, M., Fangmeier, T., Ruff, C. C., & Johnson-Laird, P. (2003). Reasoning, models, and images: Behavioral measures and cortical activity. *Journal of Cognitive Neuroscience, 15*, 559–573.

Koscik, T. R., & Tranel, D. (2012). The human ventromedial prefrontal cortex is critical for transitive inference. *Journal of Cognitive Neuroscience, 24*, 1191–1204.

Lees, R. B., & Chomsky, N. (1957). Syntactic structures. *Language, 33*, 375–408.

Li, P., & Gleitman, L. (2002). Turning the tables: Language and spatial reasoning. *Cognition, 83*, 265–294.

Locke, J. (1824). *The works of John Locke in nine volumes*. London: Rivington.
Losonsky, M. (1999). *Humboldt: 'On language': On the diversity of human language construction and its influence on the mental development of the human species*. Cambridge: Cambridge University Press.
Maruyama, M., Pallier, C., Jobert, A., Sigman, M., & Dehaene, S. (2012). The cortical representation of simple mathematical expressions. *Neuroimage, 61*, 1444–1460.
Monti, M. M., & Osherson, D. N. (2012). Logic, language and the brain. *Brain Research, 1428*, 33–42.
Monti, M. M., Osherson, D. N., Martinez, M. J., & Parsons, L. M. (2007). Functional neuroanatomy of deductive inference: A language-independent distributed network. *Neuroimage, 37*, 1005–1016.
Monti, M. M., Parsons, L. M., & Osherson, D. N. (2009). The boundaries of language and thought in deductive inference. *Proceedings of the National Academy of Sciences, 106*, 12554–12559.
Monti, M. M., Parsons, L. M., & Osherson, D. N. (2012). Thought beyond language neural dissociation of algebra and natural language. *Psychological Science, 23*(8), 914–922.
Noveck, I. A., Goel, V., & Smith, K. W. (2004). The neural basis of conditional reasoning with arbitrary content. *Cortex, 40*, 613–622.
Novick, J. M., Trueswell, J. C., & Thompson-Schill, S. L. (2010). Broca's area and language processing: Evidence for the cognitive control connection. *Language and Linguistics Compass, 4*, 906–924.
Ogawa, S., Lee, T. M., Kay, A. R., & Tank, D. W. (1990). Brain magnetic resonance imaging with contrast dependent on blood oxygenation. *Proceedings of the National Academy of Sciences, 87*(24), 9868–9872.
Osherson, D., & Falmagne, R. (1975). Logic and models of logical thinking. In R. J. Falmagne (Ed.), *Reasoning: Representation and process in children and adults* (pp. 81–92). Hillsdale, NJ: Lawrence Erlbaum Associates.
Parsons, L. M., & Osherson, D. (2001). New evidence for distinct right and left brain systems for deductive versus probabilistic reasoning. *Cerebral Cortex, 11*, 954–965.
Pinker, S. (1984). *Language learnability and language development*. Cambridge, MA: Harvard University Press.
Pinker, S., & Jackendoff, R. (2005). The faculty of language: What's special about it? *Cognition, 95*, 201–236.
Polk, T. A., & Newell, A. (1995). Deduction as verbal reasoning. *Psychological Review, 102*, 533.
Prado, J., & Noveck, I. A. (2007). Overcoming perceptual features in logical reasoning: A parametric functional magnetic resonance imaging study. *Journal of Cognitive Neuroscience, 19*, 642–657.
Prado, J., Van Der Henst, J.-B., & Noveck, I. A. (2010a). Recomposing a fragmented literature: How conditional and relational arguments engage different neural systems for deductive reasoning. *Neuroimage, 51*, 1213–1221.
Prado, J., Noveck, I. A., & Van Der Henst, J.-B. (2010b). Overlapping and distinct neural representations of numbers and verbal transitive series. *Cerebral Cortex, 20*, 720–729.
Price, C. J. (2000). The anatomy of language: Contributions from functional neuroimaging. *Journal of Anatomy, 197*, 335–359.
Reverberi, C., Cherubini, P., Frackowiak, R. S., Caltagirone, C., Paulesu, E., & Macaluso, E. (2010). Conditional and syllogistic deductive tasks dissociate functionally during premise integration. *Human Brain Mapping, 31*, 1430–1445.
Reverberi, C., Cherubini, P., Rapisarda, A., Rigamonti, E., Caltagirone, C., Frackowiak, R. S., et al. (2007). Neural basis of generation of conclusions in elementary deduction. *Neuroimage, 38*, 752–762.
Reverberi, C., Shallice, T., D'Agostini, S., Skrap, M., & Bonatti, L. L. (2009). Cortical bases of elementary deductive reasoning: Inference, memory, and metadeduction. *Neuropsychologia, 47*, 1107–1116.

Rips, L. J. (1994). *The psychology of proof: Deductive reasoning in human thinking*. Cambridge, MA: MIT Press.
Rodriguez-Moreno, D., & Hirsch, J. (2009). The dynamics of deductive reasoning: An fMRI investigation. *Neuropsychologia, 47*(4), 949–961.
Sahin, N. T., Pinker, S., & Halgren, E. (2006). Abstract grammatical processing of nouns and verbs in Broca's area: Evidence from fMRI. *Cortex, 42*, 540–562.
Sapir, E. (1929). The status of linguistics as a science. *Language, 5*, 207–214.
Sapir, E. (1931). Conceptual categories in primitive languages. *Science, 74*(1927), 578.
Spelke, E. S. (2003). What makes us smart? Core knowledge and natural language. In D. Gentner & S. Goldin-Meadow (Eds.), *Language in mind: Advances in the study of language and thought* (pp. 277–311). Cambridge, MA: MIT Press.
Spelke, E. S., & Tsivkin, S. (2001). Language and number: A bilingual training study. *Cognition, 78*, 45–88.
Stanescu-Cosson, R., Pinel, P., van de Moortele, P.-F., Le Bihan, D., Cohen, L., & Dehaene, S. (2000). Understanding dissociations in dyscalculia. *Brain, 123*, 2240–2255.
Tettamanti, M., & Weniger, D. (2006). Broca's area: A supramodal hierarchical processor? *Cortex, 42*, 491–494.
Uddén, J., & Bahlmann, J. (2012). A rostro-caudal gradient of structured sequence processing in the left inferior frontal gyrus. *Philosophical Transactions of the Royal Society of London B: Biological Sciences, 367*, 2023–2032.
Varley, R., & Siegal, M. (2000). Evidence for cognition without grammar from causal reasoning and 'theory of mind' in an agrammatic aphasic patient. *Current Biology, 10*, 723–726.
Varley, R. A., Klessinger, N. J., Romanowski, C. A., & Siegal, M. (2005). Agrammatic but numerate. *Proceedings of the National Academy of Sciences of the United States of America, 102*, 3519–3524.
Whorf, B. L. (1940). *Science and linguistics*. Indianapolis, IN: Bobbs-Merrill.
Wildgruber, D., Ackermann, H., Kreifelts, B., & Ethofer, T. (2006). Cerebral processing of linguistic and emotional prosody: fMRI studies. *Progress in Brain Research, 156*, 249–268.

Part II
Interactions Within and Between Systems

Chapter 6
Multisensory Integration in Speech Processing: Neural Mechanisms of Cross-Modal Aftereffects

Niclas Kilian-Hütten, Elia Formisano, and Jean Vroomen

6.1 Introduction

6.1.1 Multisensory Integration

Traditionally, perceptual neuroscience has focused on unimodal information processing. This is true also for investigations of speech processing, where the auditory modality was the natural focus of interest. Given the complexity of neuronal processing, this was a logical step, considering that the field was still in its infancy. However, it is clear that this restriction does not do justice to the way we perceive the world around us in everyday interactions. Very rarely is sensory information confined to one modality. Instead, we are constantly confronted with a stream of input to several or all senses and already in infancy, we match facial movements with their corresponding sounds (Campbell et al. 2001; Kuhl and Meltzoff 1982). Moreover, the information that is processed by our individual senses does not stay separated. Rather, the different channels interact and influence each other, affecting perceptual interpretations and constructions (Calvert 2001). Consequently, in the last 15–20 years, the perspective in cognitive science and perceptual neuroscience

N. Kilian-Hütten
Department of Psychiatry, Columbia University College of Physicians and Surgeons, New York, NY, USA

Department of Cognitive Neuroscience, Maastricht University, Maastricht, The Netherlands
e-mail: n.kilianhuetten@gmail.com

E. Formisano
Maastricht Brain Imaging Center, Maastricht University, Maastricht, The Netherlands
e-mail: e.formisano@maastrichtuniversity.nl

J. Vroomen (✉)
Department of Cognitive Neuropsychology, Tilburg University, Tilburg 5000 LE, The Netherlands
e-mail: J.Vroomen@uvt.nl

M. Mody (ed.), *Neural Mechanisms of Language*, Innovations in Cognitive Neuroscience, DOI 10.1007/978-1-4939-7325-5_6

has shifted to include investigations of such multimodal integrative phenomena. Facilitating cross-modal effects have consistently been demonstrated behaviorally (Shimojo and Shams 2001). When multisensory input is congruent (e.g., semantically and/or temporally) it typically lowers detection thresholds (Frassinetti et al. 2002), shortens reaction times (Forster et al. 2002; Schröger and Widmann 1998), and decreases saccadic eye movement latencies (Hughes et al. 1994) as compared to unimodal exposure. When incongruent input is (artificially) added in a second modality, this usually has opposite consequences (Sekuler et al. 1997).

6.1.2 Audiovisual Speech Perception

It becomes increasingly clear then, that in the case of spoken communication, the auditory modality is not of exclusive relevance. Indeed, visual speech signals in the form of lip movements, head movements, and gestures exert a significant influence on the perception of the auditory signal (Ross et al. 2007; Sumby and Pollack 1954; von Kriegstein 2012). It has been shown that extra-oral movements of the speaker's face and head correlate with the fundamental frequency and the amplitude of the speaker's voice (Munhall and Buchan 2004; Yehia et al. 2002) and that the addition of this information improves intelligibility (Munhall et al. 2004). However, the most informative visual aspect of speech is movement of the articulators. Over 60 years ago, it was demonstrated that processing of auditory speech signals improves when lip movements are visible (Sumby and Pollack 1954). This is especially relevant when the auditory signal is degraded or there is a substantial amount of overlapping, irrelevant signal. This can be easily appreciated in the noisy environment of a cocktail party or a poster session at a scientific conference, where one often focuses one's eyes more on the interlocutor's mouth to aid comprehension. Lip-reading is beneficial for auditory detection (Reisberg et al. 1987) and comprehension (Macleod and Summerfield 1990; Middelweerd and Plomp 1987; Sumby and Pollack 1954) and leads to improved performance equivalent to an increase in auditory signal-to-noise between 4 dB and 6 dB (Macleod and Summerfield 1990; Middelweerd and Plomp 1987) or even 12–15 dB (Sumby and Pollack 1954). It appears that these effects are present even in the absence of auditory noise (Remez 2012) and are strongest in moderate levels of noise (Ma et al. 2009; Ross et al. 2007).

Enhancement effects as discussed above reflect the most common situations: Here, auditory and visual channels are congruent and provide largely redundant information. In artificial situations, though, information sources can be put in conflict with each other: In the case of conflicting information, seen speech can actually alter both perceived location and identity of an auditory signal. In the ventriloquism effect, auditory and visual stimuli are presented synchronously, but shifted in space. This leads to a perceived displacement of the auditory source toward the visual one (Bertelson and Radeau 1981; Radeau and Bertelson 1977). The relative dominance of the individual modalities in this effect depends on the reliability (the inverse estimate of the noisiness) of each information source (Alais and Burr 2004), i.e., when the visual source is blurred, and thus poorly localized, the auditory modality dominates and influences the perceived location of the visual source. It has further been demonstrated that the ventriloquism illusion is not only effective in the spatial, but also in the temporal domain, where the perceived timing of a visual stimulus (e.g., a flash) can be biased toward an asynchronous sound (Vroomen and de Gelder 2004).

Besides perceived location or timing, conflicting inter-modal stimulation can even alter the perceived identity of auditory speech sounds. In a seminal paper that has now been cited more than 3100 times, McGurk and MacDonald (1976) demonstrated a powerful effect, in which lip-reading alters the percept of an auditory phoneme. When presenting a visual /ga/ together with an auditory /ba/, the resulting percept is /da/ (the McGurk illusion). It seems like the conflicting input results in a best-guess perceptual interpretation. The effect is extremely robust and has been a subject of intensive investigation since its discovery.

As a result of the behavioral findings discussed above, the last couple of decades has seen the emergence of a wealth of research tapping into the neural mechanisms of cross-modal integration, audiovisual speech perception and enhancement, as well as perceived-location and identity effects (see Calvert et al. 2004; Murray and Wallace 2011).

6.2 Audiovisual Speech Perception: Neural Mechanisms and Theories

6.2.1 The Modular View of Audiovisual Integration

Besides sub-cortical structures—predominantly the superior colliculus (SC)—and regions traditionally regarded as unisensory, such as early auditory and visual regions, visual area MT (middle temporal; responsible for the processing of motion), or the Fusiform Face Area (FFA), neuroimaging studies have identified a network of higher-order integrative areas that are involved in the processing of audiovisual speech. More specifically, middle and posterior superior temporal sulcus (pSTS), Broca's area, dorsolateral prefrontal cortex, superior precentral sulcus, supramarginal gyrus, angular gyrus, intraparietal sulcus (IPS), inferior frontal gyrus (IFG), and insula have all been implicated in these processes (Beauchamp et al. 2004a, b; Callan et al. 2001, 2003, 2004; Calvert et al. 1999, 2000, 2004; Capek et al. 2004; Miller and D'esposito 2005; Möttönen et al. 2004; Olson et al. 2002; Pekkola et al. 2005). At least for the homologues of pSTS and IPS, neurons have been found in non-human primates that respond to both auditory and visual stimulation (Ghazanfar and Schroeder 2006). Furthermore, some of these regions have been shown to be activated by silent speech reading. This is true for middle and posterior superior temporal sulcus (Callan et al. 2004; Ludman et al. 2000; MacSweeney et al. 2002; Skipper et al. 2005), inferior frontal gyrus and Broca's area (Campbell et al. 2001;

Ojanen et al. 2005; Watkins et al. 2003), and possibly even for primary auditory cortex (Calvert et al. 1997; Pekkola et al. 2005). Most studies reveal a left-over-right hemisphere asymmetry in activation (Capek et al. 2004).

Classical studies in cats' superior colliculi (Stein et al. 1988; Stein and Meredith 1990) first identified multisensory neurons that exhibit supra-additive firing patterns in response to multisensory, as compared to unisensory, input (AV > A + V). Early functional magnetic resonance imaging (fMRI) studies consequently searched for brain regions whose hemodynamic response (blood-oxygen level dependent response; BOLD) mimicked this activation pattern (Calvert et al. 2000). Since several studies failed to replicate successful results using this, quite strict, criterion (Beauchamp et al. 2004a, b; Beauchamp 2005; Laurienti et al. 2005; Stevenson et al. 2007), other criteria, such as a multimodal response that is stronger than the stronger one of the two unimodal responses, or inverse effectiveness (multisensory enhancement should *increase* as a function of stimulus quality degradation) have been applied. Most of these criteria have received a substantial amount of criticism. For a discussion of this, see Laurienti et al. (2005), James and Stevenson (2012), and Stein et al. (2009).

In spite of the criticism, left pSTS has been repeatedly and robustly implicated in audiovisual integration on the basis of these criteria (Beauchamp et al. 2004a, b; Calvert et al. 2000; Miller and D'esposito 2005; Nath et al. 2011; Stevenson and James 2009; Wright et al. 2003). Furthermore, when audiovisual stimuli are incongruent, usually a depression of activity in this region results (Campbell and Capek 2008; Wright et al. 2003) and, as mentioned before, left pSTS is activated by both audiovisual speech and by silent speech-reading (Callan et al. 2004; Campbell and Capek 2008; Capek et al. 2004; Hall et al. 2005; MacSweeney et al. 2002; Skipper et al. 2005) and differences in left STS activation have been related to language comprehension (McGettigan et al. 2012). More recently, results from electrocorticography suggest a dissociation between anterior and posterior STG in responses to audiovisual speech with clear vs noisy auditory component. The pSTG not aSTG appears to be important for multisensory integration of noisy auditory and visual speech (Ozker et al. 2017). Also, single-pulse transcranial magnetic stimulation (TMS) over pSTS has been shown to disrupt the perception of McGurk effects in a time window from 100 ms before the onset of the auditory stimulus to 100 ms after onset (Beauchamp et al. 2010).

6.2.2 Multisensory Processing as the Default Mode of Speech Perception

The findings discussed so far follow the traditional modular view that assumes that multisensory integration takes place only in higher-order multisensory regions after extensive unisensory processing in the respective cortices. However, more recently, this view has been challenged (Ghazanfar and Schroeder 2006; Schroeder et al. 2008). Increasing evidence is accumulating for the idea that multisensory processing can be regarded as the default mode of speech perception and that the integration of auditory and visual speech signals already occurs in the earliest stage of processing in, presumptively unisensory cortical areas (Driver and Noesselt 2008; Ghazanfar and Schroeder 2006; Ghazanfar 2012; Rosenblum et al. 2005). Studies in non-human primates (Ghazanfar et al. 2005, 2008; Kayser et al. 2005, 2007; Lakatos et al. 2007), as well as in humans (Besle et al. 2004, 2009; Pekkola et al. 2005; Stekelenburg and Vroomen 2007; Van Wassenhove et al. 2005; Vroomen and Stekelenburg 2010) have demonstrated the integrative influence of visual (including speech signals) and somatosensory signals on auditory processing in primary and lateral belt auditory cortex. It has been suggested that the underlying mechanism of such cross-modal modulation of early auditory cortical processing may be based on a predictive resetting of the phase of the ongoing oscillatory cycles of neuronal ensembles (Schroeder et al. 2008).

Several accounts for these early cross-modal influences on presumptively unisensory cortices exist, which offer different, although not necessarily mutually exclusive, explanations for the origin of these modulations (Driver and Noesselt 2008; Schroeder et al. 2003). More specifically, it is conceivable that (a) all cortical regions are essentially multisensory and receive input from thalamocortical pathways and direct cortico-cortical connections between different sensory cortices (Ghazanfar and Schroeder 2006); (b) that there is still a separation between unisensory and multisensory integration areas, but new convergence zones exist earlier in the hierarchy and closer to unisensory regions than previously assumed (Beauchamp et al. 2004a, b); and (c) that cross-modal modulations of sensory-specific cortical processing reflects feedback influences from multisensory convergence zones (Driver and Noesselt 2008; Jiang et al. 2001). While account (b) basically amounts to a new parcellation of cortex within the traditional perspective, account (a) represents a rather extreme new view on cortical processing. Support for account (a) comes from neuroanatomical studies demonstrating the involvement of thalamocortical (Cappe et al. 2009; Hackett et al. 2007; Lakatos et al. 2007) and monosynaptic cortico-cortical connections between primary auditory and primary visual cortex (Clavagnier et al. 2004; Falchier et al. 2002, 2010), from direct connections between voice- and face-processing areas (Blank et al. 2011), and from reports of very early post-stimulus cross-modal influences on the event-related potential (ERP; within approximately 50 ms) (Giard and Peronnet 1999; Molholm et al. 2002; Senkowski et al. 2007). However, feedback connections from convergence zones, such as pSTS, still seem to clearly outnumber direct connections between early sensory-specific cortices (Falchier et al. 2002). Also, it is unclear whether the information transmitted along this route is stimulus- or percept-specific and, thus, reflects actual multisensory integration, or whether it represents more general modulations, such as attention or arousal effects (Driver and Noesselt 2008). Furthermore, ERP studies demonstrating extremely early effects based on the additive model have been criticized for not taking into account common, non-specific activity, such as attention effects and anticipatory, and motor, responses (Cappe et al. 2010; Gondan and Röder 2006; Teder-Sälejärvi et al. 2002). Controlling for these factors typically delays the effects to approximately 60–100 ms. Over the last decade it has been shown that non-linear multisensory interactions in the ERP follow from topographic modulations, and result in sub-additive responses that are functionally coupled within primary auditory and visual cortices, as well as pSTS (Cappe et al. 2010). Support for account (c) the notion that multisensory effects in sensory-specific cortices reflect feedback influences from higher-order convergence zones comes from comparisons of latencies in the electroencephalogram (EEG) (Besle et al. 2004; Ponton et al. 2009), from fMRI studies investigating functional contrasts (Calvert et al. 2000), connectivity (Noesselt et al. 2007) and experimental differentiations between integration responses and perceptual effects (Kilian-Hütten et al. 2011a, b; Sohoglu et al. 2012), and from studies interfering with normal brain functioning in order to establish cause-and-effect relationships in cats (Jiang et al. 2001) and humans (Beauchamp et al. 2010).

It has become increasingly clear that the different accounts just discussed are not mutually exclusive, but that feedforward, lateral, and feedback connections exist and that multisensory integration involves all of these, relying more or less on particular types of integration depending on stimulus and task context (Besle et al. 2008, 2009; Driver and Noesselt 2008; Schroeder et al. 2003).

6.2.3 Theoretical Accounts of Audiovisual Speech Perception

In the context of audiovisual speech perception, one important distinction here might be what has been termed correlation versus complementary mode (Campbell and Capek 2008). These two proposed modes of processing (correlation versus complementary) focus on the relation between auditory and visual speech stimuli and are based on the observation that auditory comprehension benefits from visual information in two ways; first, the auditory and the visual signal are highly correlated in terms of temporal dynamics and the speech processing system exploits these redundancies (correlation mode), and second, when the quality of the auditory signal is degraded, or certain speech segments are acoustically ambiguous, the visual signal can help disambiguate the acoustics and aid perception (complementary mode) (Campbell and Capek 2008). There is support for the idea that the specific locus of multisensory integration is affected by the relative importance of these modes in a particular experimental context. Callan et al. (2004) varied the visibility of facial detail using spatial filtering and could show that, while middle temporal gyrus (MTG) activation was increased when fine detail was accessible, pSTS activation was unaffected, lending evidence to the idea that pSTS is driven primarily by the dynamic aspects of the audiovisual speech stream (correlated mode), rather than by specific facial detail (complementary mode) (Campbell and Capek 2008).

The differences between correlation mode and complementary mode may reflect a more general phenomenon; reliability-based cue weighting (Fetsch et al. 2012; Nath and Beauchamp 2011; Sheppard et al. 2013). It has been demonstrated repeatedly in behavioral studies that, in multisensory integration, subjects give stronger weighting to the more reliable modality and that these weightings are adapted in a dynamic, context-dependent fashion (Alais and Burr 2004; Ernst and Banks 2002;

Ma et al. 2009). This leads to an optimal solution, because it creates estimates with the lowest possible variance, which, in turn, results in superior perceptual performance as compared to what can be achieved based on either unisensory signal alone, or with any predetermined weighting pattern (Fetsch et al. 2012). In audiovisual speech perception, this may result in dynamic changes of functional connectivity between auditory and visual sensory cortices, respectively, and integration cortices, such as pSTS, depending on the reliability of the auditory and visual speech signals (Nath and Beauchamp 2011).

In the light of all these findings, it is vital from a computational perspective to understand how auditory recognition can benefit from visual input. One framework that may be applicable here is based on a theory that has been proposed to more generally account for perceptual inference (i.e., the recognition of perceived objects) and perceptual learning effects (Friston 2005). This scheme, referred to commonly as "predictive coding," rests on the idea that an integral part in perceptual inference is minimizing free energy, or more pragmatically, error. This is done by relying on a hierarchical model, where sensory responses at lower levels of the hierarchy are predicted at higher levels. In return, lower levels send prediction errors to higher levels, enabling learning. In other words, this idea relies on an empirical Bayesian model, where prior expectations can be formed, which in turn, exert their influence in the light of sensory evidence in a dynamic and context-dependent fashion. In the realm of multisensory integration, priors may originate in another modality (von Kriegstein 2012). For speech, it is conceivable that visual information (lip movement) biases processing in the auditory cortex, probably (but not necessarily) via feedback connections from higher-order convergence zones. Recently, support for this idea has been found in purely auditory speech (Clos et al. 2014), in auditory speech primed by written text (Sohoglu et al. 2012), and in audiovisual speech (Arnal et al. 2011; Noppeney et al. 2008). It has been suggested that, in an oscillation framework, higher frequencies (in the gamma (~30–60 Hz) and high-gamma (~70–80 Hz) frequency range) may be primarily involved in the signaling of bottom-up prediction errors, while slower frequencies (beta frequency range) would communicate top-down predictions (Arnal et al. 2011; Arnal and Giraud 2012).

6.3 Cross-Modal Aftereffects

6.3.1 Hysteresis Versus Adaptation

The discussion so far has focused on situations where the speech signals from both modalities (auditory and visual) are presented concurrently and affect each other directly. This is true for normal speech perception and also when one or both modalities are noisy, or when the perceived location and/or identity of a stimulus are altered on the fly. However, there are also cases where cross-modal effects can alter unisensory perception beyond the immediate presentation in time. Such aftereffects have been found in unisensory phenomena, such as the waterfall illusion (Purkinje 1820) where after looking at a waterfall for an extended period of time, stationary rocks, for instance, seem to be moving upward, or the prism experiments by Stratton (1897) in which he found that a radical conflict between proprioception and visual field (which was turned upside down using special goggles) over time led to an adaptation of visual perception. Aftereffects can have two directions: Negative aftereffects like the waterfall illusion, the tilt aftereffect (the perceived change in orientation of a line or grating after prolonged exposure to another oriented line or grating; Gibson and Radner 1937), or color-opponency likely reflect a “fatigue” of neural sensors, while positive aftereffects, such as prism adaptation, likely reflect a “learning” of new sensory arrangements. Such attractive aftereffects (making a similar percept more likely) are also known as hysteresis, while repulsive aftereffects (making a similar percept less likely) are also referred to as adaptation.

In the multisensory domain, the ventriloquist illusion has been shown to produce attractive aftereffects after prolonged exposure (Bertelson et al. 2006; Radeau and Bertelson 1974, 1977). As was mentioned before, in this effect, the perceived location of a sound is shifted in space toward a synchronously presented visual target (Bermant and Welch 1976; Bertelson and Radeau 1981; Bertelson and Aschersleben 1998; Klemm 1909). The associated aftereffects are in line with this immediate effect—the perceived location of sounds presented in isolation after audiovisual exposure is shifted toward the visual stimuli presented during the exposure phase. In audiovisual speech perception, traditionally, contrastive effects have been found, evident in selective speech adaptation (Roberts and Summerfield 1981), where the repeated presentation of a nonambiguous phoneme leads to a decrease of the probability of reporting the same percept when tested with an ambiguous phoneme. In other words, repeated exposure to a nonambiguous /aba/ leads to a reduction of subsequent /aba/ perception, a phenomenon that, as mentioned before, may be explained by neuronal fatigue (Anstis et al. 1998; Eimas and Corbit 1973) (but see: Diehl et al. 1978; Diehl 1981; Samuel 1986).

6.3.2 The Initial Study on Audiovisual Recalibration of Auditory Speech Perception

Bertelson et al. (2003), however, were inspired by the findings of hysteresis effects in the ventriloquist illusion and investigated the possibility of similar aftereffects in the domain of audiovisual speech perception. They hypothesized that the crucial manipulation for achieving a hysteresis effect, as opposed to an adaptation effect, was the ambiguity of the auditory component of the adapter stimuli. While in the classical McGurk effect, as well as in selective speech adaptation paradigms, a nonambiguous phoneme is used, Bertelson et al. (2003) synthesized an ambiguous phoneme halfway between a nonambiguous /aba/ and a nonambiguous /ada/ (*A?*) and dubbed this sound onto videos of a speaker pronouncing /aba/ (A?Vb) or /ada/

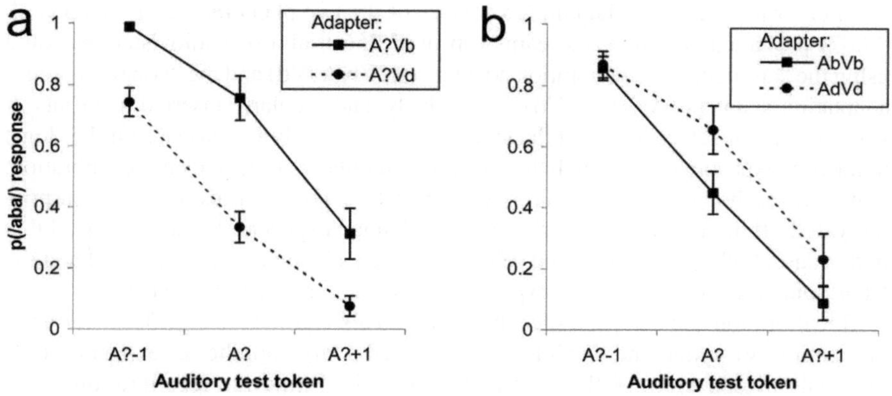

Fig. 6.1 The graphs show the proportion of /aba/ judgments after exposure to an adapter consisting of (**a**) the participant's ambiguous auditory token (A?) combined with either visual /aba/ (A?Vb) or visual /ada/ (A?Vd) or (**b**) a nonambiguous auditory token (Ab or Ad) combined with the congruent visual stimulus, /aba/ (AbVb) or /ada/ (AdVd). Figure (**a**) clearly shows a recalibration effect, while (**b**) indicates selective speech adaptation

(A?Vd), respectively. They showed that the repeated exposure to one type of video (A?Vb or A?Vd, respectively) increased the probability of corresponding perceptual interpretations in auditory-only post-tests. In the original study, eight videos were presented, followed by six post-tests (A? twice, plus the two tokens closest to it on the /aba/-/ada/ continuum; A? − 1 and A? + 1). This procedure was repeated a large number of times in random order. It could be shown that the proportion of /aba/ responses was significantly higher following A?Vb exposure than following A?Vd exposure (see Fig. 6.1a).

Crucially, when nonambiguous auditory tokens were dubbed onto the videos (AbVb/AdVd), the contrasting effect was found, i.e., selective speech adaptation (Fig. 6.1b). This is especially remarkable because subjects could not perceptually distinguish the ambiguous (A?Vb/A?Vd) from the nonambiguous (AbVb/AdVd) videos (Vroomen et al. 2004). This finding also rules out the possibility of an explicit strategy endorsed by the subjects, since they could not know, for a given block, whether they were exposed to ambiguous or nonambiguous stimuli.

6.3.3 Differences Between Recalibration and Selective Speech Adaptation

Besides the disparity in direction of effect, recalibration, and selective speech adaptation also differ in a number of psychophysical characteristics; namely in buildup, dissipation, and the necessity of processing stimuli within a "speech mode" (Vroomen and Baart 2012).

In order to investigate the buildup courses of the two phenomena, Vroomen et al. (2007) presented continuous streams (up until 256 trials) of audiovisual exposure using the "ambiguous" recalibration adapters (A?Vb/A?Vd) and the "nonambiguous" adaptation adapters (AbVb/AdVd), respectively, and regularly inserted test trials. It was shown that selective speech adaptation effects linearly increased with the (log-) number of exposure trials, which fits with an accumulative fatigue idea. Recalibration, however, reached ceiling level already after eight exposure trials and then, surprisingly, fell off after 32 exposure trials with prolonged exposure. It was suggested that in this case, both recalibration and selective speech adaptation run in parallel and the latter dominates with prolonged exposure due to an increasing effect size.

Recalibration and selective speech adaptation also turned out to differ in terms of dissipation. Vroomen and de Gelder (2004) used a large number of exposure trials (50 trials of one kind), followed by 60 test trials. While the recalibration effect already dissipated after as little as six exposure trials, the effects of selective speech adaptation were still manifest even after 60 trials.

Lastly, Vroomen and Baart (2009) investigated the speech-specificity of both effects. In order to do so, they relied on sine-wave speech, a manipulation of speech stimuli which reduces the richness of the speech sound by removing all of its natural attributes and retaining only the pattern of vocal tract resonance changes; hence, these stimuli can be perceived either as speech or as non-speech, depending on the subject's perceptual mode. In order to manipulate perceptual mode, subjects were trained to distinguish two sine-wave stimuli as either /omso/ and /onso/ (speech mode) or as stimulus 1 and 2 (non-speech mode). It could be shown that recalibration crucially relies on being in speech mode (recalibration took place in speech mode, but not in non-speech mode), while perceptual mode had no effect on selective speech adaptation (which was effective in both modes).

6.4 The Neural Mechanisms of Cross-Modal Recalibration

The uniqueness of cross-modal recalibration makes it an intriguing subject for neuroscientific investigation. The reason for this is twofold; first, the aforementioned psychophysical differences suggested that cross-modal recalibration has a distinct neural substrate from that of selective speech adaptation (i.e., neuronal fatigue), rendering it crucial to find the origin of this perceptual biasing signal. Second, the recalibration paradigm yielded the unique possibility of comparing the neuronal responses underlying the differential perceptual interpretation (/ada/ or /aba/) of physically identical, auditory stimuli. In other words, recalibration made it possible to experimentally disentangle acoustics from perception and to examine the neuronal underpinnings of a purely perceptual difference.

In order to investigate these aspects of the recalibration phenomenon, Kilian-Hütten et al. (2011a, b) adapted the classical setup employed in the original Bertelson et al. (2003) study for the functional magnetic resonance imaging (fMRI) environment (Fig. 6.2). Subjects were presented with blocks of eight identical audiovisual

Fig. 6.2 Schematic overview of the paradigm used by Kilian-Hütten et al. (2011a, b), which was based on the behavioral study by Bertelson et al. (2003). Each run consisted of ten mini runs, which each comprised eight audiovisual exposure trials (A?Vd or A?Vb), followed by six auditory post-test trials. Audiovisual exposure was presented following a block design, while an event-related presentation scheme was applied for the auditory test trials

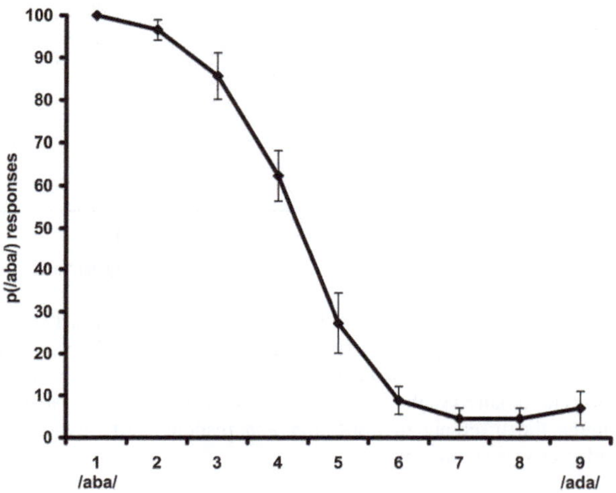

Fig. 6.3 Results of the auditory pretest. The mean proportion (p) of /aba/ classifications across the 11 participants for each stimulus on the nine-step /aba/-/ada/continuum are presented. Sound #4 was chosen as A? for eight of the participants and sound #5 for the remaining three

adapters (A?Vb and A?Vd, respectively), which were each followed by six auditory test trials consisting of forced choice perceptual judgments (/aba/ vs. /ada/) of ambiguous stimuli (A?, A? + 1, A? − 1). The most ambiguous stimulus on the nine-step continuum ranging from /aba/ to /ada/ was identified individually per subject in a pretest (Fig. 6.3). Behaviorally, the results of the original psychophysical study (Bertelson et al 2003) could be replicated (Fig. 6.4). Scanning was performed using a mixed block/event-related design, where audiovisual exposure trials were presented in blocks, while auditory test trials were presented in slow event-related fashion. This enabled the authors to achieve high signal-to-noise ratios for the exposure blocks, while preserving the possibility of analyzing auditory test trials on a trial-by-trial basis, depending on the subjects' perceptual judgments. This last point was crucial in order to allow for the possibility of investigating the neural substrate of the purely perceptually distinct categorization (/aba/ vs. /ada/) of physically identical stimuli (A?).

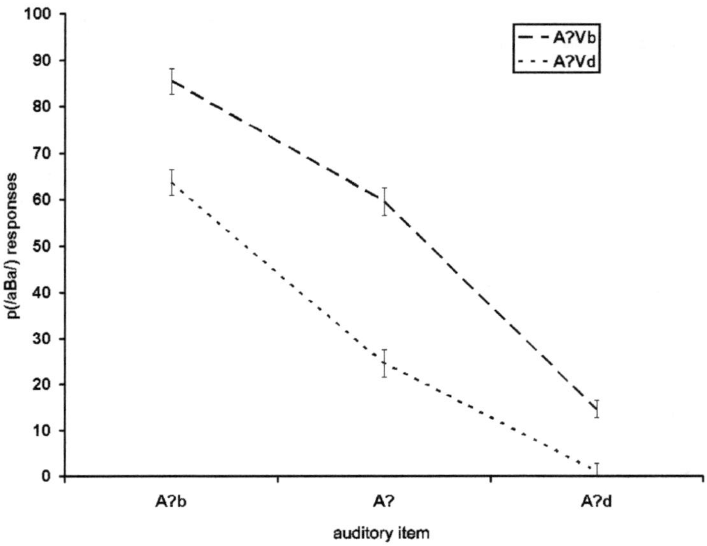

Fig. 6.4 Behavioral results of the auditory post-tests in Kilian-Hütten et al. (2011a, b). For the participant's ambiguous auditory item and its two neighbors on the continuum, the graph shows the proportion of /aba/ classifications after exposure to an audiovisual adapter comprised of the ambiguous item paired with either visual /aba/ (A?Vb) or with visual /ada/ (A?Vd). For all three auditory items, the difference in proportion /aba/ responses after exposure to the A?Vb adapter vs. the A?Vd adapter was significant

6.4.1 The Effects of Recalibration: Decoding Perceptual Interpretations of Ambiguous Phonemes

Traditionally, neuroimaging studies on perception have faced a confounding issue. This is because distinct percepts usually follow distinct physical stimuli. In other words, when /aba/ is presented, a subject perceives /aba/ and when /ada/ is presented, a subject perceives /ada/. Hence, when applying subtraction logic, as is traditional procedure in fMRI research, the two conditions that are being compared usually differ not only in percept, but also in physical stimulus characteristics. When one is interested in the neural substrate of perception proper, and not of stimulus processing per se, this is a problem. The recalibration phenomenon, together with event-related stimulus presentation, allowed disentangling auditory perception from stimulus acoustics.

Kilian-Hütten et al. (2011a) combined the recalibration paradigm with a machine learning approach in order to decode auditory perception on a trial-by-trial basis from the fMRI signal. This approach, also referred to as multivoxel pattern analysis (MVPA), applies machine learning methods to the multivariate analysis of fMRI data sets (Formisano et al. 2008; Haxby et al. 2001; Haynes and Rees 2005a, b). A pattern classification algorithm is typically trained, using a large set of trials, to associate a given experimental condition or cognitive state with a distributed pattern of fMRI responses. The trained classifier can then be tested on new, unseen fMRI patterns to decode the associated cognitive state. In other words, rather than predicting brain responses from experimental conditions, as in the general linear model (GLM; a regression analysis using regressors based on the timing of experimental conditions), MVPA can "predict" the experimental condition from the brain response, which is why it has often been denoted as a "brain reading" approach. One particular class of MVPA algorithms are support vector machines (SVM).

Kilian-Hütten et al. (2011a) trained an SVM to learn the association between multivariate patterns of fMRI signal and corresponding labels, determined by the subjects' perceptual interpretations. In other words, while the physical stimuli were identical for both labels, the percept differed, and the SVM was trained to decode the respective percept on a trial-by-trial basis from the fMRI signal. The analysis was anatomically confined to the temporal lobes in order to test the hypothesis that abstract auditory representations can be found in early auditory cortex. Besides accuracy levels (which were significantly above chance, as validated with permutation testing), it is interesting to visualize the spatial activation patterns that were used for classification. To this end, group discriminative maps, i.e., maps of the cortical locations that contributed most to the discrimination of conditions, were created after cortex-based alignment (Goebel et al. 2006) of single-subject discriminative maps (Staeren et al. 2009). For SVM analyses it is meaningful, at an individual map level, to rank the features (i.e., voxels) relatively according to their contribution to the discrimination. In the resulting group-level discriminative maps, a cortical location (vertex) was color-coded if it was present among the 30% of most discriminative vertices in the corresponding individual discriminative maps of at least seven of the 11 subjects. As can be seen in Fig. 6.5, these maps identified left-lateralized clusters of vertices along the posterior bank of Heschl's gyrus, Heschl's sulcus, and, adjacently, in the anterior portion of planum temporale (PT). Additional clusters of smaller extent were found at the left temporoparietal junction and, bilaterally, on middle superior temporal gyrus (STG) and sulcus (STS).

These results showed that pure perceptual interpretation of physically identical phonemes can be decoded from activation patterns in early auditory cortex. Thus, beyond the basic acoustic analysis of sounds, constructive perceptual information is present in regions within the anterior PT, tangent to the posterior bank of Heschl's gyrus and sulcus.

6.4.2 The Origin of Recalibration: Predicting Recalibration Strength from Cortical Activation

The results just discussed above concentrated on the effects of cross-modal recalibration, i.e., the change in auditory perception. The obvious next question that arose was: Where does cross-modal recalibration itself take place and, thus, what is the

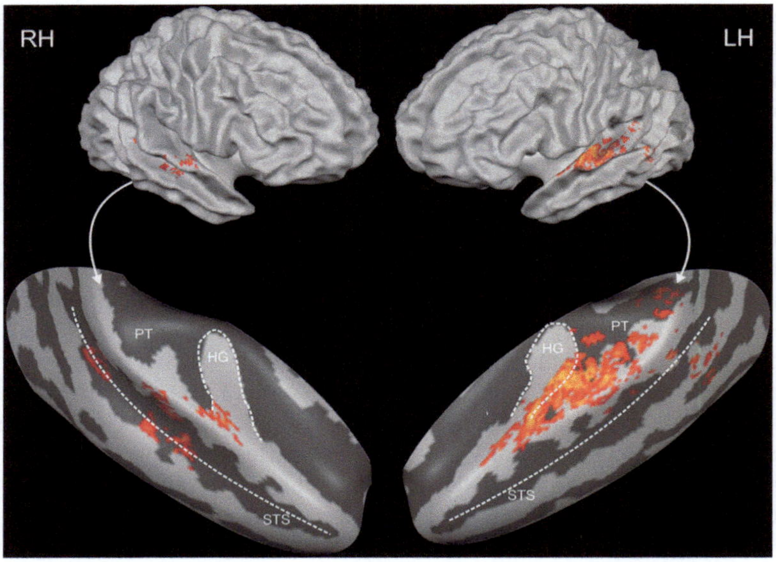

Fig. 6.5 Discriminative map from Kilian-Hütten et al. (2011a). Group map of the 30% of active voxels most discriminative for the purely perceptual difference between /aba/ and /ada/. A location was color-coded if it was present on the individual maps of at least seven of the 11 subjects. Maps are overlaid on the reconstructions of the average hemispheres of the 11 subjects (*top*) and on inflated reconstructions of the *right* and *left* temporal lobes of these average hemispheres (*bottom*). *RH* right hemisphere, *LH* left hemisphere, *HG* Heschl's gyrus

origin of the perceptually biasing effect? In order to answer these research questions, Kilian-Hütten et al. (2011b) focused their efforts on cortical activation during the audiovisual exposure trials. In a first step, a simple comparison between blocks of audiovisual exposure and baseline identified a network of brain regions corresponding to those generally found in audiovisual speech perception paradigms (primary and extrastriate visual areas, primary and early auditory areas, STG/STS, inferior frontal sulcus (IFS), premotor cortex, and inferior parietal lobe, touching upon angular and supermarginal gyri and the intraparietal sulcus). This was expected and replicated prior results. However, this does not mean that this whole network is responsible for the recalibration effect. In order to identify the subset of regions for which this is indeed true, the authors applied a behaviorally defined contrast. The strength of the recalibration effect is variable from one given exposure block to the next and can be quantified as the number of auditory post-tests perceived in line with the type of exposure block (A?Vb or A?Vd). These values could be employed to identify brain regions whose activation during the exposure blocks varied with the strength of the recalibration effect. The cortical activation in these areas, thus, predicted the perceptual tendency later in time. Beyond the basic identification of responsive regions, recalibration, thus, made it possible to pinpoint those regions which were responsive to audiovisual stimulation and which, further, were functionally relevant in driving the biasing perceptual effect. The network of regions for

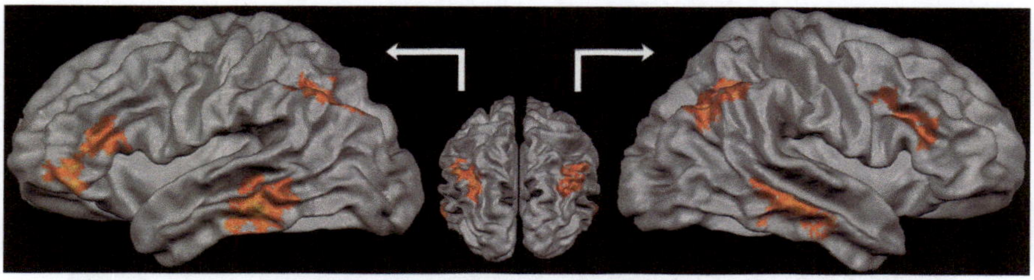

Fig. 6.6 Group results for the behaviorally weighted contrast used by Kilian-Hütten et al. (2011b) overlaid on the average hemispheres obtained from the cortex-based alignment procedure. Shown are bilateral IPL, IFS, and posterior middle temporal gyrus

which this was true included bilateral IPL, IFS, and posterior middle temporal gyrus (Fig. 6.6). These results were further corroborated by a functional connectivity analysis (psychophysiological interaction analysis; PPI), which demonstrated that this network of areas exhibits increased functional/effective connectivity with the left auditory cortex during blocks of audiovisual exposure relative to baseline.

These findings are in correspondence with the results from other investigations of perceptual learning (Myers et al. 2009; Naumer et al. 2009; Raizada and Poldrack 2007). Gilbert et al. (2001, p. 681) define perceptual learning as "improving one's ability, with practice, to discriminate differences in the attributes of simple stimuli." In the case of recalibration, the disambiguating information from audiovisual exposure biases auditory perception such that it can be regarded as improved in reference to the (momentary) demands of sensory reality. What seems to happen in the case of cross-modal recalibration, thus, is that integrative audiovisual learning effects take place in the identified network, which in turn affect later constructive (auditory) perceptual processes.

6.4.3 Theories of Audiovisual Speech Perception and the Neural Substrate of Recalibration

Taking together the results from the studies discussed above, a full neural model of cross-modal recalibration emerges (Fig. 6.7). A higher-order network including IPL, IFS, and MTG is suggested to process integrative learning effects, and consequently install a perceptual bias in auditory regions, most prominently the left Heschl's sulcus and the planum temporale, influencing future constructive auditory perception.

This interpretation of the results is in line with a model of the neural mechanisms of hysteresis and adaptation recently put forward by Schwiedrzik et al. (2014). In their exclusively visual study, they were able to dissociate hysteresis and adaptation effects in a single paradigm. What they found was that, while adaptation effects were largely confined to early visual areas, hysteresis effects mapped onto a more widespread and higher-order fronto-parietal network. Using a modeling approach, they showed that their results could be explained in a Bayesian framework.

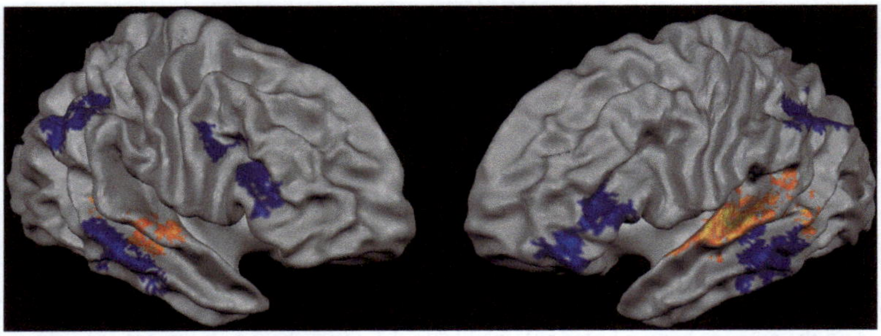

Fig. 6.7 The proposed model (Kilian-Hütten et al. 2011a, b). A higher-order network (in *blue*) including IPL, IFS, and MTG is suggested to process integrative learning effects (cross-modal recalibration), and consequently install a perceptual bias (the prior) in auditory regions (in *red* and *orange*), most prominently the left Heschl's sulcus and the planum temporale, influencing future perceptual interpretations of sensory input (the likelihood function), resulting in recalibrated auditory perception

A Bayesian approach is well-suited for this problem, because it takes into account both the available evidence (the likelihood function) and knowledge about the world (the prior). In terms of hysteresis and adaptation, this means that perception is determined by the sensory evidence (which is bimodal in the case of bistable stimuli) and by the prior. While the former is influenced by adaptation (neuronal fatigue), the latter is determined by hysteresis. More specifically, when a certain interpretation is primed, the prior is adjusted toward this interpretation, inducing hysteresis. This can be regarded as a special case of the general model of predictive coding. As discussed before, in this framework, sensory responses at lower levels of the hierarchy are predicted at higher levels. In return, lower levels send prediction errors to higher levels, enabling learning. Also Bayesian in nature, this model states that prior expectations can be formed and, in turn, exert their influence in the light of sensory evidence in a dynamic- and context-dependent fashion.

In the case of cross-modal recalibration, this would mean that the fronto-parietal network computes the prior and communicates the outcome to early auditory regions. Here, a representation of the perceptual interpretation can be decoded. This representation is almost exclusively determined by the prior, because the ambiguity of the stimulus creates a bimodal likelihood function, which has to be disambiguated by the prior before perception can take place.

6.5 Conclusion

Audiovisual speech perception involves the coordinated processing of a large network of brain regions. Besides unisensory cortices, higher-order regions are involved in the convergence and integration of multisensory input, as well as in the semantic and cognitive appraisal of this information. Cross-talk between the senses has an impact on perception, including perceived location and perceived identity of sensory input. Multisensory processing improves recognition accuracy particularly for speech in noise through multiple stages on integration supported by distinct neuroanatomical mechanisms (Peelle and Sommers 2015). It seems that in order to optimally integrate auditory and visual information in the perception of speech, the brain exploits several connections, including feedforward, lateral, and feedback pathways. The relative importance of these connections depends on stimulus and task conditions. One important factor in this context is the relative reliability of the sensory input, as explained by reliability-based cue weighting and predictive coding frameworks.

These frameworks are also important in accounting for the neural bases of cross-modal perceptual aftereffects, such as recalibration. Recalibration is a hysteresis effect in that it elicits post-exposure perceptual biases that are in line with the percepts experienced during audiovisual exposure. This, along with differences in buildup, dissipation, and the necessity of processing stimuli in "speech mode," suggests that the neural substrate of this effect is different from that of selective speech adaptation (neuronal fatigue). Using a machine learning approach and behaviorally weighted GLM contrasts, it could be shown that cross-modal recalibration is reflected in integrative audiovisual learning effects that take place in a higher-order network involving IPL, IFS, and posterior middle temporal gyrus and which then install a perceptual bias in early auditory regions, impacting later auditory perception.

These results can be interpreted along the lines of a Bayesian framework, closely related to reliability-based cue weighting and predictive coding. Following this rationale, the fronto-parietal network would compute a Bayesian prior and communicate the outcome to early auditory regions. Here, perception is determined on the basis of this prior and incoming sensory evidence (the likelihood function). Since, in the case of recalibration, sensory evidence is ambiguous, perception is mostly determined by the prior. Hence, perception of the A? stimuli follows the audiovisual exposure blocks- /ada/ for A?Vd and /aba/ for A?Vb.

In a separate study, a modeling approach applied to the data from Vroomen et al. (2007) showed that a Bayesian model could explain the behavioral data reflecting both the phenomena of phonetic recalibration and selective speech adaptation (Kleinschmidt and Jaeger 2011). The authors suggest that this "belief-updating model" could provide a unified explanation for both phenomena. The results discussed before, however, demonstrate that the neural underpinnings of both phenomena appear to be distinct, suggesting that separate mechanisms are at play. In future, it will be essential to devise a single coherent model that can explain, both, the phenomena of cross-modal recalibration and selective speech adaptation, while taking their neural mechanisms into account, i.e., an ecologically valid Bayesian model of phonemic aftereffects.

References

Alais, D., & Burr, D. (2004). The ventriloquist effect results from near-optimal bimodal integration. *Current Biology, 14*, 257–262.

Anstis, S., Verstraten, F. A., & Mather, G. (1998). The motion aftereffect. *Trends in Cognitive Sciences, 2*, 111–117.

Arnal, L. H., & Giraud, A. L. (2012). Cortical oscillations and sensory predictions. *Trends in Cognitive Sciences, 16*, 390–398.
Arnal, L. H., Wyart, V., & Giraud, A.-L. (2011). Transitions in neural oscillations reflect prediction errors generated in audiovisual speech. *Nature Neuroscience, 14*, 797–801.
Beauchamp, M. S. (2005). Statistical criteria in FMRI studies of multisensory integration. *Neuroinformatics, 3*, 93–113.
Beauchamp, M. S., Argall, B. D., Bodurka, J., Duyn, J. H., & Martin, A. (2004a). Unraveling multisensory integration: Patchy organization within human STS multisensory cortex. *Nature Neuroscience, 7*, 1190–1192.
Beauchamp, M. S., Lee, K. E., Argall, B. D., & Martin, A. (2004b). Integration of auditory and visual information about objects in superior temporal sulcus. *Neuron, 41*, 809–823.
Beauchamp, M. S., Nath, A. R., & Pasalar, S. (2010). fMRI-guided transcranial magnetic stimulation reveals that the superior temporal sulcus is a cortical locus of the McGurk effect. *Journal of Neuroscience, 30*, 2414–2417.
Bermant, R. I., & Welch, R. B. (1976). Effect of degree of separation of visual-auditory stimulus and eye position upon spatial interaction of vision and audition. *Perceptual and Motor Skills, 43*, 487–493.
Bertelson, P., & Aschersleben, G. (1998). Automatic visual bias of perceived auditory location. *Psychonomic Bulletin & Review, 5*, 482–489.
Bertelson, P., Frissen, I., Vroomen, J., & De Gelder, B. (2006). The aftereffects of ventriloquism: Patterns of spatial generalization. *Attention, Perception, & Psychophysics, 68*, 428–436.
Bertelson, P., & Radeau, M. (1981). Cross-modal bias and perceptual fusion with auditory-visual spatial discordance. *Attention, Perception, & Psychophysics, 29*, 578–584.
Bertelson, P., Vroomen, J. & De Gelder, B. (2003). Visual recalibration of auditory speech identification: A McGurk aftereffect. *Psychological Sciences*, 14, 592–597.
Besle, J., Bertrand, O., & Giard, M.-H. (2009). Electrophysiological (EEG, sEEG, MEG) evidence for multiple audiovisual interactions in the human auditory cortex. *Hearing Research, 258*, 143–151.
Besle, J., Fischer, C., Bidet-Caulet, A., Lecaignard, F., Bertrand, O., & Giard, M.-H. (2008). Visual activation and audiovisual interactions in the auditory cortex during speech perception: Intracranial recordings in humans. *Journal of Neuroscience, 28*, 14301–14310.
Besle, J., Fort, A., Delpuech, C., & Giard, M. H. (2004). Bimodal speech: Early suppressive visual effects in human auditory cortex. *European Journal of Neuroscience, 20*, 2225–2234.
Blank, H., Anwander, A., & von Kriegstein, K. (2011). Direct structural connections between voice-and face-recognition areas. *Journal of Neuroscience, 31*, 12906–12915.
Callan, D. E., Callan, A. M., Kroos, C., & Vatikiotis-Bateson, E. (2001). Multimodal contribution to speech perception revealed by independent component analysis: A single-sweep EEG case study. *Cognitive Brain Research, 10*, 349–353.
Callan, D. E., Jones, J. A., Munhall, K., Callan, A. M., Kroos, C., & Vatikiotis-Bateson, E. (2003). Neural processes underlying perceptual enhancement by visual speech gestures. *Neuroreport, 14*, 2213–2218.
Callan, D. E., Jones, J. A., Munhall, K., Kroos, C., Callan, A. M., & Vatikiotis-Bateson, E. (2004). Multisensory integration sites identified by perception of spatial wavelet filtered visual speech gesture information. *Journal of Cognitive Neuroscience, 16*, 805–816.
Calvert, G., Spence, C., & Stein, B. E. (Eds.) (2004). *The handbook of multisensory processes*. Cambridge, MA: MIT Press.
Calvert, G. A. (2001). Crossmodal processing in the human brain: Insights from functional neuroimaging studies. *Cerebral Cortex, 11*, 1110–1123.
Calvert, G. A., Brammer, M. J., Bullmore, E. T., Campbell, R., Iversen, S. D., & David, A. S. (1999). Response amplification in sensory-specific cortices during crossmodal binding. *Neuroreport, 10*, 2619–2623.
Calvert, G. A., Campbell, R., & Brammer, M. J. (2000). Evidence from functional magnetic resonance imaging of crossmodal binding in the human heteromodal cortex. *Current Biology, 10*, 649–657.

Calvert, G. A., Bullmore, E. T., Brammer, M. J., Campbell, R., Williams, S. C., McGuire, P. K., et al. (1997). Activation of auditory cortex during silent lipreading. *Science, 276*, 593–596.

Campbell, R., & Capek, C. (2008). Seeing speech and seeing sign: Insights from a fMRI study. *International Journal of Audiology, 47*, S3–S9.

Campbell, R., MacSweeney, M., Surguladze, S., Calvert, G., McGuire, P., Suckling, J., et al. (2001). Cortical substrates for the perception of face actions: An fMRI study of the specificity of activation for seen speech and for meaningless lower-face acts (gurning). *Cognitive Brain Research, 12*, 233–243.

Capek, C. M., Bavelier, D., Corina, D., Newman, A. J., Jezzard, P., & Neville, H. J. (2004). The cortical organization of audio-visual sentence comprehension: An fMRI study at 4 Tesla. *Cognitive Brain Research, 20*, 111–119.

Cappe, C., Rouiller, E. M., & Barone, P. (2009). Multisensory anatomical pathways. *Hearing Research, 258*, 28–36.

Cappe, C., Thut, G., Romei, V., & Murray, M. M. (2010). Auditory–visual multisensory interactions in humans: Timing, topography, directionality, and sources. *Journal of Neuroscience, 30*, 12572–12580.

Clavagnier, S., Falchier, A., & Kennedy, H. (2004). Long-distance feedback projections to area V1: Implications for multisensory integration, spatial awareness, and visual consciousness. *Cognitive, Affective, & Behavioral Neuroscience, 4*, 117–126.

Clos, M., Langner, R., Meyer, M., Oechslin, M. S., Zilles, K., & Eickhoff, S. B. (2012). Effects of prior information on decoding degraded speech: An fMRI study. *Human Brain Mapping, 35*, 61–74.

Diehl, R. L. (1981). Feature detectors for speech: A critical reappraisal. *Psychological Bulletin, 89*, 1.

Diehl, R. L., Elman, J. L., & McCusker, S. B. (1978). Contrast effects on stop consonant identification. *Journal of Experimental Psychology: Human Perception and Performance, 4*, 599.

Driver, J., & Noesselt, T. (2008). Multisensory interplay reveals crossmodal influences on 'sensory-specific' brain regions, neural responses, and judgments. *Neuron, 57*, 11–23.

Eimas, P. D., & Corbit, J. D. (1973). Selective adaptation of linguistic feature detectors. *Cognitive Psychology, 4*, 99–109.

Ernst, M. O., & Banks, M. S. (2002). Humans integrate visual and haptic information in a statistically optimal fashion. *Nature, 415*, 429–433.

Falchier, A., Clavagnier, S., Barone, P., & Kennedy, H. (2002). Anatomical evidence of multimodal integration in primate striate cortex. *Journal of Neuroscience, 22*, 5749–5759.

Falchier, A., Schroeder, C. E., Hackett, T. A., Lakatos, P., Nascimento-Silva, S., Ulbert, I., et al. (2010). Projection from visual areas V2 and prostriata to caudal auditory cortex in the monkey. *Cerebral Cortex, 20*, 1529–1538.

Fetsch, C. R., Pouget, A., DeAngelis, G. C., & Angelaki, D. E. (2012). Neural correlates of reliability-based cue weighting during multisensory integration. *Nature Neuroscience, 15*, 146–154.

Formisano, E., De Martino, F., Bonte, M., & Goebel, R. (2008). "Who" is saying "what"? Brain-based decoding of human voice and speech. *Science, 322*, 970–973.

Forster, B., Cavina-Pratesi, C., Aglioti, S. M., & Berlucchi, G. (2002). Redundant target effect and intersensory facilitation from visual-tactile interactions in simple reaction time. *Experimental Brain Research, 143*, 480–487.

Frassinetti, F., Bolognini, N., & Làdavas, E. (2002). Enhancement of visual perception by cross-modal visuo-auditory interaction. *Experimental Brain Research, 147*, 332–343.

Friston, K. (2005). A theory of cortical responses. *Philosophical Transactions of the Royal Society of London B: Biological Sciences, 360*, 815–836.

Ghazanfar, A. A. (2012). Unity of the senses for primate vocal communication. In M. T. Murray & M. M. Wallace (Eds.), *The neural bases of multisensory processes* (pp. 653–666). Boca Raton: CRC.

Ghazanfar, A. A., Chandrasekaran, C., & Logothetis, N. K. (2008). Interactions between the superior temporal sulcus and auditory cortex mediate dynamic face/voice integration in rhesus monkeys. *Journal of Neuroscience, 28*, 4457–4469.

Ghazanfar, A. A., Maier, J. X., Hoffman, K. L., & Logothetis, N. K. (2005). Multisensory integration of dynamic faces and voices in rhesus monkey auditory cortex. *Journal of Neuroscience, 25*, 5004–5012.
Ghazanfar, A. A., & Schroeder, C. E. (2006). Is neocortex essentially multisensory? *Trends in Cognitive Sciences, 10*, 278–285.
Giard, M. H., & Peronnet, F. (1999). Auditory-visual integration during multimodal object recognition in humans: A behavioral and electrophysiological study. *Journal of Cognitive Neuroscience, 11*, 473–490.
Gibson, J. J., & Radner, M. (1937). Adaptation, after-effect and contrast in the perception of tilted lines. I. Quantitative studies. *Journal of Experimental Psychology, 20*, 453.
Gilbert, C. D., Sigman, M., & Crist, R. E. (2001). The neural basis of perceptual learning. *Neuron, 31*, 681–697.
Goebel, R., Esposito, F., & Formisano, E. (2006). Analysis of functional image analysis contest (FIAC) data with brainvoyager QX: From single-subject to cortically aligned group general linear model analysis and self-organizing group independent component analysis. *Human Brain Mapping, 27*, 392–401.
Gondan, M., & Röder, B. (2006). A new method for detecting interactions between the senses in event-related potentials. *Brain Research, 1073*, 389–397.
Hackett, T. A., De La Mothe, L. A., Ulbert, I., Karmos, G., Smiley, J., & Schroeder, C. E. (2007). Multisensory convergence in auditory cortex. II. Thalamocortical connections of the caudal superior temporal plane. *Journal of Comparative Neurology, 502*, 924–952.
Hall, D. A., Fussell, C., & Summerfield, A. Q. (2005). Reading fluent speech from talking faces: Typical brain networks and individual differences. *Journal of Cognitive Neuroscience, 17*, 939–953.
Haxby, J. V., Gobbini, M. I., Furey, M. L., Ishai, A., Schouten, J. L., & Pietrini, P. (2001). Distributed and overlapping representations of faces and objects in ventral temporal cortex. *Science, 293*, 2425–2430.
Haynes, J.-D., & Rees, G. (2005a). Predicting the orientation of invisible stimuli from activity in human primary visual cortex. *Nature Neuroscience, 8*, 686–691.
Haynes, J.-D., & Rees, G. (2005b). Predicting the stream of consciousness from activity in human visual cortex. *Current Biology, 15*, 1301–1307.
Hughes, H. C., Reuter-Lorenz, P. A., Nozawa, G., & Fendrich, R. (1994). Visual-auditory interactions in sensorimotor processing: Saccades versus manual responses. *Journal of Experimental Psychology: Human Perception and Performance, 20*, 131–153.
Jiang, W., Wallace, M. T., Jiang, H., Vaughan, J. W., & Stein, B. E. (2001). Two cortical areas mediate multisensory integration in superior colliculus neurons. *Journal of Neurophysiology, 85*, 506–522.
Kayser, C., Petkov, C. I., Augath, M., & Logothetis, N. K. (2005). Integration of touch and sound in auditory cortex. *Neuron, 48*, 373–384.
Kayser, J., Tenke, C. E., Gates, N. A., & Bruder, G. E. (2007). Reference-independent ERP old/new effects of auditory and visual word recognition memory: Joint extraction of stimulus-and response-locked neuronal generator patterns. *Psychophysiology, 44*, 949–967.
Kilian-Hütten, N., Valente, G., Vroomen, J., & Formisano, E. (2011a). Auditory cortex encodes the perceptual interpretation of ambiguous sound. *Journal of Neuroscience, 31*, 1715–1720.
Kilian-Hütten, N., Vroomen, J., & Formisano, E. (2011b). Brain activation during audiovisual exposure anticipates future perception of ambiguous speech. *Neuroimage, 57*, 1601–1607.
Kleinschmidt, D., & Jaeger, T. F. (2011). A Bayesian belief updating model of phonetic recalibration and selective adaptation. In *Proceedings of the 2nd Workshop on Cognitive Modeling and Computational Linguistics* (pp. 10–19). Association for Computational Linguistics.
Klemm, O. (1909). Lokalisation von Sinneseindrücken bei disparaten Nebenreizen. [Localization of sensory impressions with disparate distractors]. *Psychologische Studien (Wundt) 5*, 73–161.
Kuhl, P. K., & Meltzoff, A. N. (1982). *The bimodal perception of speech in infancy*. Washington, DC: American Association for the Advancement of Science.
Lakatos, P., Chen, C.-M., O'Connell, M. N., Mills, A., & Schroeder, C. E. (2007). Neuronal oscillations and multisensory interaction in primary auditory cortex. *Neuron, 53*, 279–292.

Laurienti, P. J., Perrault, T. J., Stanford, T. R., Wallace, M. T., & Stein, B. E. (2005). On the use of superadditivity as a metric for characterizing multisensory integration in functional neuroimaging studies. *Experimental Brain Research, 166*, 289–297.

Ludman, C., Summerfield, A. Q., Hall, D., Elliott, M., Foster, J., Hykin, J. L., et al. (2000). Lip-reading ability and patterns of cortical activation studied using fMRI. *British Journal of Audiology, 34*, 225–230.

Ma, W. J., Zhou, X., Ross, L. A., Foxe, J. J., & Parra, L. C. (2009). Lip-reading aids word recognition most in moderate noise: A Bayesian explanation using high-dimensional feature space. *PLoS One, 4*, e4638.

Macleod, A., & Summerfield, Q. (1990). A procedure for measuring auditory and audiovisual speech-reception thresholds for sentences in noise: Rationale, evaluation, and recommendations for use. *British Journal of Audiology, 24*, 29–43.

MacSweeney, M., Woll, B., Campbell, R., McGuire, P. K., David, A. S., Williams, S. C., et al. (2002). Neural systems underlying British Sign Language and audio-visual English processing in native users. *Brain, 125*, 1583–1593.

McGettigan, C., Faulkner, A., Altarelli, I., Obleser, J., Baverstock, H., & Scott, S. K. (2012). Speech comprehension aided by multiple modalities: Behavioural and neural interactions. *Neuropsychologia, 50*, 762–776.

McGurk, H., & MacDonald, J. (1976). Hearing lips and seeing voices. *Nature, 264*, 746–748

Middelweerd, M., & Plomp, R. (1987). The effect of speechreading on the speech-reception threshold of sentences in noise. *The Journal of the Acoustical Society of America, 82*, 2145–2147.

Miller, L. M., & D'esposito, M. (2005). Perceptual fusion and stimulus coincidence in the cross-modal integration of speech. *Journal of Neuroscience, 25*, 5884–5893.

Molholm, S., Ritter, W., Murray, M. M., Javitt, D. C., Schroeder, C. E., & Foxe, J. J. (2002). Multisensory auditory–visual interactions during early sensory processing in humans: A high-density electrical mapping study. *Cognitive Brain Research, 14*, 115–128.

Möttönen, R., Schürmann, M., & Sams, M. (2004). Time course of multisensory interactions during audiovisual speech perception in humans: A magnetoencephalographic study. *Neuroscience Letters, 363*, 112–115.

Munhall, K. G., & Buchan, J. N. (2004). Something in the way she moves. *Trends in Cognitive Science, 8*, 51–53.

Munhall, K. G., Jones, J. A., Callan, D. E., Kuratate, T., & Vatikiotis-Bateson, E. (2004). Visual prosody and speech intelligibility head movement improves auditory speech perception. *Psychological Science, 15*, 133–137.

Murray, M. T., & Wallace, M. M. (Eds.) (2011). *The neural bases of multisensory processes*. Boca Raton: CRC.

Myers, E. B., Blumstein, S. E., Walsh, E., & Eliassen, J. (2009). Inferior frontal regions underlie the perception of phonetic category invariance. *Psychological Science, 20*, 895–903.

Nath, A. R., & Beauchamp, M. S. (2011). Dynamic changes in superior temporal sulcus connectivity during perception of noisy audiovisual speech. *Journal of Neuroscience, 31*, 1704–1714.

Nath, A. R., Fava, E. E., & Beauchamp, M. S. (2011). Neural correlates of interindividual differences in children's audiovisual speech perception. *Journal of Neuroscience, 31*, 13963–13971.

Naumer, M. J., Doehrmann, O., Müller, N. G., Muckli, L., Kaiser, J., & Hein, G. (2009). Cortical plasticity of audio–visual object representations. *Cerebral Cortex, 19*, 1641–1653.

Noesselt, T., Rieger, J. W., Schoenfeld, M. A., Kanowski, M., Hinrichs, H., Heinze, H.-J., et al. (2007). Audiovisual temporal correspondence modulates human multisensory superior temporal sulcus plus primary sensory cortices. *Journal of Neuroscience, 27*, 11431–11441.

Noppeney, U., Josephs, O., Hocking, J., Price, C. J., & Friston, K. J. (2008). The effect of prior visual information on recognition of speech and sounds. *Cerebral Cortex, 18*, 598–609.

Ojanen, V., Möttönen, R., Pekkola, J., Jääskeläinen, I. P., Joensuu, R., Autti, T., et al. (2005). Processing of audiovisual speech in Broca's area. *Neuroimage, 25*, 333–338.

Olson, I. R., Gatenby, J. C., & Gore, J. C. (2002). A comparison of bound and unbound audio–visual information processing in the human cerebral cortex. *Cognitive Brain Research, 14*, 129–138.

Ozker, M., Schepers, I. M., Magnotti, J. F., Yosher, D., & Beauchamps, M. (2017). A double dissociation between anterior and posterior superior temporal gyrus for processing audiovi-

sual speech demonstrated by electrocorticography. *Journal of Cognitive Neuroscience, 2916*, 1044–1060.
Pekkola, J., Ojanen, V., Autti, T., Jääskeläinen, I. P., Möttönen, R., Tarkiainen, A., et al. (2005). Primary auditory cortex activation by visual speech: An fMRI study at 3 T. *Neuroreport, 16*, 125–128.
Peelle, J.E. & Sommers, M.S. (2015). Prediction and constraint in audiovisual speech perception. *Cortex, 68*, 169–181.
Ponton, C. W., Bernstein, L. E., & Auer, E. T. (2009). Mismatch negativity with visual-only and audiovisual speech. *Brain Topography, 21*, 207–215.
Purkinje, J. (1820). Beiträge zur näheren Kenntnis des Schwindels. *Med Jahrb kuk Staates (Wien), 6*, 23–35.
Radeau, M., & Bertelson, P. (1974). The after-effects of ventriloquism. *The Quarterly Journal of Experimental Psychology, 26*, 63–71.
Radeau, M., & Bertelson, P. (1977). Adaptation to auditory-visual discordance and ventriloquism in semirealistic situations. *Perception & Psychophysics, 22*, 137–146.
Raizada, R. D., & Poldrack, R. A. (2007). Selective amplification of stimulus differences during categorical processing of speech. *Neuron, 56*, 726–740.
Reisberg, D., McLean, J., & Goldfield, A. (1987). Easy to hear but hard to understand: A lip-reading advantage with intact auditory stimuli. In B. Dodd & R. Campbell (Eds.), *Hearing by eye: The psychology of lip-reading* (pp. 97–113). London: Erlbaum.
Remez, R. E. (2012). Three puzzles of multimodal speech perception. In G. Bailly, P. Perrier, & E. Vatikiotis-Bateson (Eds.), *Audiovisual speech processing* (pp. 4–20). Cambridge: Cambridge University Press.
Roberts, M., & Summerfield, Q. (1981). Audiovisual presentation demonstrates that selective adaptation in speech perception is purely auditory. *Attention, Perception, & Psychophysics, 30*, 309–314.
Rosenblum, L. D., Pisoni, D., & Remez, R. (2005). Primacy of multimodal speech perception. In D. Pisoni & R. Remez (Eds.) *Handbook of speech perception* (pp. 51–78). Malden: Blackwell.
Ross, L. A., Saint-Amour, D., Leavitt, V. M., Javitt, D. C., & Foxe, J. J. (2007). Do you see what I am saying? Exploring visual enhancement of speech comprehension in noisy environments. *Cerebral Cortex, 17*, 1147–1153.
Samuel, A. G. (1986). Red herring detectors and speech perception: In defense of selective adaptation. *Cognitive Psychology, 18*, 452–499.
Schroeder, C. E., Lakatos, P., Kajikawa, Y., Partan, S., & Puce, A. (2008). Neuronal oscillations and visual amplification of speech. *Trends in Cognitive Sciences, 12*, 106–113.
Schroeder, C. E., Smiley, J., Fu, K. G., McGinnis, T., O'Connell, M. N., & Hackett, T. A. (2003). Anatomical mechanisms and functional implications of multisensory convergence in early cortical processing. *International Journal of Psychophysiology, 50*, 5–17.
Schröger, E., & Widmann, A. (1998). Speeded responses to audiovisual signal changes result from bimodal integration. *Psychophysiology, 35*, 755–759.
Schwiedrzik, C. M., Ruff, C. C., Lazar, A., Leitner, F. C., Singer, W., & Melloni, L. (2014). Untangling perceptual memory: Hysteresis and adaptation map into separate cortical networks. *Cerebral Cortex, 24*, 1152–1164.
Sekuler, R., Sekuler, A. B., & Lau, R. (1997). Sound alters visual motion perception. *Nature, 385*, 308.
Senkowski, D., Talsma, D., Grigutsch, M., Herrmann, C. S., & Woldorff, M. G. (2007). Good times for multisensory integration: Effects of the precision of temporal synchrony as revealed by gamma-band oscillations. *Neuropsychologia, 45*, 561–571.
Sheppard, J. P., Raposo, D., & Churchland, A. K. (2013). Dynamic weighting of multisensory stimuli shapes decision-making in rats and humans. *Journal of Vision, 13*, 4.
Shimojo, S., & Shams, L. (2001). Sensory modalities are not separate modalities: Plasticity and interactions. *Current Opinion in Neurobiology, 11*, 505–509.
Skipper, J. I., Nusbaum, H. C., & Small, S. L. (2005). Listening to talking faces: Motor cortical activation during speech perception. *Neuroimage, 25*, 76–89.
Sohoglu, E., Peelle, J. E., Carlyon, R. P., & Davis, M. H. (2012). Predictive top-down integration of prior knowledge during speech perception. *Journal of Neuroscience, 32*, 8443–8453.
Staeren, N., Renvall, H., De Martino, F., Goebel, R., & Formisano, E. (2009). Sound categories are represented as distributed patterns in the human auditory cortex. *Current Biology, 19*, 498–502.

Stein, B., & Meredith, M. (1990). Multimodal integration: Neural and behavioral solutions for dealing with stimuli from different modalities. *Annals of the New York Academy of Science, 606*, 51–70.

Stein, B. E., Huneycutt, W. S., & Meredith, M. A. (1988). Neurons and behavior: The same rules of multisensory integration apply. *Brain Research, 448*, 355–358.

Stein, B. E., Stanford, T. R., Ramachandran, R., Perrault, T. J., & Rowland, B. A. (2009). Challenges in quantifying multisensory integration: Alternative criteria, models, and inverse effectiveness. *Experimental Brain Research, 198*, 113.

Stekelenburg, J. J., & Vroomen, J. (2007). Neural correlates of multisensory integration of ecologically valid audiovisual events. *Journal of Cognitive Neuroscience, 19*, 1964–1973.

Stevenson, R. A., Geoghegan, M. L., & James, T. W. (2007). Superadditive BOLD activation in superior temporal sulcus with threshold non-speech objects. *Experimental Brain Research, 179*, 85–95.

Stevenson, R. A., & James, T. W. (2009). Audiovisual integration in human superior temporal sulcus: Inverse effectiveness and the neural processing of speech and object recognition. *Neuroimage, 44*, 1210–1223.

Stratton, G. M. (1897). Vision without inversion of the retinal image. *Psychological Review, 4*, 341.

Sumby, W. H., & Pollack, I. (1954). Visual contribution to speech intelligibility in noise. *The Journal of the Acoustical Society of America, 26*, 212–215.

Teder-Sälejärvi, W., McDonald, J., Di Russo, F., & Hillyard, S. (2002). An analysis of audio-visual crossmodal integration by means of event-related potential (ERP) recordings. *Cognitive Brain Research, 14*, 106–114.

Van Wassenhove, V., Grant, K. W., & Poeppel, D. (2005). Visual speech speeds up the neural processing of auditory speech. *Proceedings of the National Academy of Sciences of the United States of America, 102*, 1181–1186.

von Kriegstein, K. (2012). A multisensory perspective on human auditory communication. In M. T. Murray & M. M. Wallace (Eds.), *The neural bases of multisensory processes* (pp. 683–702). Boca Raton: CRC.

Vroomen, J., & Baart, M. (2009). Phonetic recalibration only occurs in speech mode. *Cognition, 110*, 254–259.

Vroomen, J., & Baart, M. (2012). Phonetic recalibration in audiovisual speech. In M. T. Murray & M. M. Wallace (Eds.), *The neural bases of multisensory processes* (pp. 363–380). Boca Raton: CRC.

Vroomen, J., & de Gelder, B. (2004). Temporal ventriloquism: Sound modulates the flash-lag effect. *Journal of Experimental Psychology: Human Perception and Performance, 30*, 513–518.

Vroomen, J., & Stekelenburg, J. J. (2010). Visual anticipatory information modulates multisensory interactions of artificial audiovisual stimuli. *Journal of Cognitive Neuroscience, 22*, 1583–1596.

Vroomen, J., van Linden, S., De Gelder, B., & Bertelson, P. (2007). Visual recalibration and selective adaptation in auditory–visual speech perception: Contrasting build-up courses. *Neuropsychologia, 45*, 572–577.

Vroomen, J., van Linden, S., Keetels, M., De Gelder, B., & Bertelson, P. (2004). Selective adaptation and recalibration of auditory speech by lipread information: Dissipation. *Speech Communication, 44*, 55–61.

Watkins, K. E., Strafella, A. P., & Paus, T. (2003). Seeing and hearing speech excites the motor system involved in speech production. *Neuropsychologia, 41*, 989–994.

Wright, T. M., Pelphrey, K. A., Allison, T., McKeown, M. J., & McCarthy, G. (2003). Polysensory interactions along lateral temporal regions evoked by audiovisual speech. *Cerebral Cortex, 13*, 1034–1043.

Yehia, H. C., Kuratate, T., & Vatikiotis-Bateson, E. (2002). Linking facial animation, head motion and speech acoustics. *Journal of Phonetics, 30*, 555–568.

Chapter 7
A Neuro-cognitive View of the Bilingual Brain

Katherine Midgley

Neuro-cognitive research dedicated to bilingual[1] language processing aims to provide a deep descriptive understanding of when and where bilingual language processing happens in the brain including how the bilingual case differs from, as well as how it resembles, monolingual language processing. The intention is not to convey any sense of hierarchy between monolingual and bilingual processing, but rather to highlight the importance of both these processes in research on language comprehension.

Much of the current knowledge about how humans use and comprehend language comes from studies involving participants from one specific language group and addresses single-language questions. This approach has been extremely helpful in advancing our knowledge of the neural substrates of language processing in general. It is from this position that the field has taken its first thrust for the most part. Almost in retort to this monolingual position research articles dealing with questions of bilingual language processing frequently begin with the observation that in our modern, globally connected world, bilingualism is the norm rather than the exception. Another related observation is that bilinguals are also monolinguals: that is to say that they can function in a monolingual mode therefore any knowledge of language processing in general feeds the domain of bilingual language processing.

Although linguists and psychologists have been interested in bilingual processing for many years, limitations in approaches to studying language use in actual language users hindered progress for a long time. This is because while there are some overtly observable indices of language processing in the case of language production, language comprehension is not directly observable so subtle differences in processing are very difficult to discern. Moreover, language processing occurs very rapidly in normal production and comprehension (~3 words per second)

[1] Bilingual throughout this chapter is understood to mean bilingual or multilingual.

K. Midgley (✉)
Department of Psychology, San Diego State University, San Diego, CA 92182-4211, USA
e-mail: kmidgley@sdsu.edu

M. Mody (ed.), *Neural Mechanisms of Language*, Innovations in Cognitive Neuroscience, DOI 10.1007/978-1-4939-7325-5_7

placing heavy demands on any measure of processing. We will return to the ramifications of this aspect of language shortly.

Over the past few decades there has been a substantial growth in the number of studies of bilingual language processing as well as an increase in the variety of methods and models used to describe bilingual language processing. And with the increased recognition that a full understanding of language processing must include a description of its neural underpinnings, one recent and burgeoning approach to studying bilingual language processing has involved the use of neuro-cognitive measures. As used here, neuro-cognitive implies a measure that is directly sensitive to the underlying neuronal basis of perceptual, cognitive, and linguistic functions. Therefore, it would seem appropriate to review some of the literature on bilingual language processing that uses such neuro-cognitive measures. We focus on studies that use electroencephalography (EEG) given the exquisite temporal sensitivity of this method, making it ideal for capturing the millisecond timescale of language processing. We draw on findings that address five interrelated issues with regard to how words are processed in the bilingual brain: language selectivity, language control, language interactivity, language differences, and second language acquisition. The results from these studies highlight the contribution of priming paradigms in combination with electrophysiological measures like event-related potentials (ERP) to our understanding of bilingual processing in the brain.

A note—this chapter will not attempt a comprehensive review of neuro-cognitive studies of bilingualism or its development, but rather will focus on a subset of the literature concerned with what we would argue is the most fundamental component of bilingual language use—the processing of words in one or both of a bilingual's two languages. For those interested in a discussion of neuro-cognitive studies of language production or sentence processing we refer the reader to reviews by Kotz (2009) and Abutalebi and Green (2007).

The study of the mental and neural mechanisms involved in recognizing words has enjoyed a rich tradition in cognitive science in general, but has arguably played an even more central role in studies of bilingual language processing. Over the last 35 years, the amount of research dedicated to this topic has increased substantially (see another recent review by van Heuven and Dijkstra (2010)) and the data collected in these studies has been instrumental in the formulation and testing of a number of influential models of bilingual language processing. Below we briefly discuss the strengths and weaknesses of a number of modern neuro-cognitive measures that have been used to study word processing in bilinguals. This is followed by a more focused review of studies using one of these techniques.

7.1 Measuring Brain Activity

A growing variety of neuro-cognitive techniques have been used to study bilingual language processing as well as other cognitive processes that have been proposed to be impacted by bilingualism. An increasingly common method, functional magnetic resonance imaging (fMRI) uses a measure of blood flow in the brain. Changes in regional blood flow (viz., the BOLD response) as a function of levels of an experimentally manipulated variable are taken as evidence that a particular brain area(s) is involved in the specific aspect of bilingual language processing that was manipulated by the investigator.

Perhaps the greatest strength of blood flow measures is their good spatial resolution; they can be used to pinpoint, in the range of a few square millimeters in the case of fMRI, the location of the language processes of interest within the intact brain. However, with this advantage of precision in space comes what is arguably a big limitation—blood flow dependent measures lack temporal precision—especially in the time range critical for decomposing language processes such as word processing. Listeners and skilled readers encounter roughly two to four words per second in typical language environments such as normal conversation or during reading. This means that all of the mental/neural operations that go into processing each word happen in something less than half a second. It stands to reason that to untangle the sensory, perceptual, and linguistic processes involved in word recognition requires the use of experimental techniques with a temporal sensitivity much shorter than half a second. The bottom line is that the bulk of the processes of interest to bilingual language researchers occur on a timescale substantially shorter than 2 s, the sensitivity of measures of changes in blood flow.

In addition to a lack of temporal specificity, blood flow measures also tend to lag substantially (on the order of seconds) behind the processes of interest (the blood flow response is said to be *sluggish*), making the information that can be discerned more of an "offline" measure of processing. Note that this is the same problem that many behavioral measures of processing suffer from—the response monitored occurs after the process of interest which leaves open questions about whether other later processes/brain systems might be involved in a particular response.

This weakness in *temporal resolution*, however, is not a problem for electrophysiological measures which directly tap some aspect of the electrical activity of the brain and which can be recorded with millisecond resolution (see Luck 2014, for a tutorial on these techniques). The most widely used measures of this type include the raw electroencephalogram (EEG) and its time-locked averaged variant—event-related potentials (ERPs), though more recently time-frequency analysis of EEG and magnetoencephalography (MEG; the magnetic counterpart of EEG) signals are providing a novel view of how de-synchronized and synchronized brain oscillatory activity may guide language processing (see Chap. 4, Fellner and Hanslmayr). All of these measures can be recorded in a non-invasive fashion from sensors placed outside the body, which like fMRI and fNIRS, make them widely amenable for use in non-clinical populations. It is also possible to record electrical signals in humans from the surface of the brain and even deep within certain brain structures. However, due to the invasive nature of these approaches such studies can only be attempted in clinical populations already undergoing some type of invasive brain procedure. Needless to say, such studies in a bilingual context are rare.

7.1.1 EEG and Cognitive Processing

Electrophysiological recording techniques are thought to reflect rapidly changing post-synaptic potentials generated by populations of interacting neurons in the cerebral cortex (Nunez and Srinivasan 2006) which explains their excellent temporal resolution and their sensitivity to changes in the neural signature of sensory, perceptual, cognitive, and linguistic processes. In other words, these measures meet the criteria of true online indices of processing. Moreover, because of the continuous nature of the electrical/magnetic activity it is possible to obtain ongoing measures starting from the moment a linguistic event begins and extending all the way through hundreds of milliseconds of processing.

As with all dependent variables, electrophysiological measures also have their limitations. From a neuro-cognitive perspective, the biggest problem is that when used alone these measures lack spatial resolution. So, while it is possible to record EEG and MEG from sensors spaced across the scalp, which provides information about the configuration of neural sources, this spatial information lacks precision in the sense that it is not possible to determine the actual location within the brain of the electrical events recorded. This is because of what is referred to as the inverse problem which basically states that from electrophysiological measures recorded outside the brain, it is impossible to back-track to the exact location of their origin—that is, to a unique set of neural sources (from a mathematical characterization the problem is ill posed, see Nunez and Srinivasan 2006). This limitation of electrophysiological measures would seem to disqualify them as useful methods for examining the neuro-cognition of bilingualism. However, the exquisite temporal information of the technique combined with the ability to reveal something about the configuration of neural sources (in the form of differences in scalp distribution patterns resulting from recording from multiple sensors) and the relative ease and inexpensiveness with which data can be collected has resulted in the technique being widely used.

More recently researchers have proposed using a multimodal approach whereby data from more than one neuroimaging method are combined to overcome the limitations of any one measure (e.g., Leonard et al. 2010). While combining techniques is challenging, many in this field believe that a true neuro-cognitive accounting of language (and bilingual) processing will only be achieved through such efforts. Twenty years of bilingual research using the above techniques has now resulted in a literature of several hundred studies. To keep this chapter manageable we will restrict this review to a subset of those studies—specifically those that have used electrophysiological measures. Our rationale for this approach is twofold. First, as discussed above, these techniques are arguably the best suited of the neuro-cognitive methods for unraveling the time course of bilingual word processing which we consider as key to yielding important insights into second language processing. Second, because we ourselves use this approach in our research we are best equipped to describe studies in this area (the reader interested in other neuro-cognitive methods in bilingual research should consult several recent reviews of these methodologies—e.g., Hernandez 2009). We will further narrow our focus to the examination of studies that have explored bilingual word processing, and in particular will concentrate on studies that have addressed five interrelated issues with regard to how words are processed in the bilingual brain: language selectivity, language control, language interactivity, language differences, and second language acquisition.

7.2 Interrelated Issues in Bilingual Processing

7.2.1 Language Selectivity

A key question in research on bilingualism is one of storage, access, and related processing of the different languages in a speaker's repertoire. Bilinguals are capable of effortlessly speaking and listening (and reading and writing) in one of their languages without obvious intrusions from the other, non-relevant language. They can also handle rapid switches from one language to the other, whether in production or comprehension. Bilinguals can translate—but they are not forever translating at every moment. Bilinguals must keep their languages separate, but not so separate that switching is overly demanding or translation intractable.

We focus on the nature of lexical storage and access in this section. If access is not language selective, it follows that access is to a common language store. If access is selective it could be to two independent stores. We will see that there is an abundance of data supporting one integrated lexicon for bilinguals (as opposed to two independent stores, one for each language) and supporting non-selective access to an integrated lexicon. Though there is now a consensus on this point, the question was explored and debated in the past.

Beginning with behavioral studies, the topic of single vs. independent lexicons was elegantly investigated by van Heuven et al. (1998). The authors sought evidence for activation of lexical items (words) from the non-relevant language during reading of a list of words in the relevant language; otherwise stated is there activation of words from language B while processing strictly in language A. Any evidence for the activation of items from language B would support a nonselective account of bilingual word processing (and also support an integrated lexicon account). The authors manipulated orthographic neighborhood counts (neighborhood density) to explore this question; that is they manipulated cross-language orthographic neighborhood counts. In a monolingual context orthographic neighbors of a word are the other words of the language that have the same length as a target word but differ in only one letter. In a bilingual context orthographic neighbors can belong to the same language as the word being processed (language A) or to the non-relevant language (language B). Van Heuven et al., working with Dutch–English bilinguals, showed that the number of both, relevant and non-relevant language neighbors influenced word processing in first language (L1) and second language (L2). It would seem that this type of result could only occur if lexical items in both languages are partially activated simultaneously and suggests that such items are contained in a single integrated lexicon. One problem with this interpretation is that when using behavioral measures to assess language processing the response typically occurs substantially after the putative perceptual-linguistic process of interest. This leaves open the possibility that the observed effects could be due to influences after the process of interest and in this sense, like behavioral measures, fMRI lacks temporal precision and clear interpretability.

Neighborhood density has also been found to influence ERPs. Recall that ERPs, unlike behavioral measures and fMRI, can reveal differences in processing throughout the time course of word processing. Effects occurring in different time windows can therefore be related to different subcomponents of language processing. The N400, a negative-going waveform typically peaking around 400 ms, is perhaps the most ubiquitous ERP component reported in language studies. And because the story of the N400 is so germane to many of the topics we will be covering from here on, we will take a slight detour at this point for a brief review of this ERP component.

N400

The N400 was first reported by Kutas and Hillyard (1980). Their landmark study compared sentence ending words that were anomalous in the context of the earlier part of the sentence (e.g., He takes his coffee with cream and *dog*) compared to final words that fit the context (e.g., He takes his coffee with cream and *sugar*). Anomalous words (*dog*) produced a large negative-going ERP component that peaked near 400 ms post-word onset, but had a time course starting around 300 ms and extending to 500 ms (by time course we mean the time interval where the ERPs for two conditions differ). In subsequent studies a very similar N400 effect was demonstrated even without a sentence context. For example, using a single prime word Holcomb (1988) showed that the ERP response to a target word produced a larger N400 if the prime was semantically unrelated to the target (e.g., car-*DOG*) compared to when it was related (e.g., cat-*DOG*—see Fig. 7.1). These results along with others (Van Petten and Kutas 1991) suggest that words in general generate this ERP response and that the N400 can be used as that long sought after marker for semantic processing.

Importantly, other studies have clearly demonstrated that the N400 is not language specific. For example, pictures of objects in a priming context have also been shown to modulate the N400 (Holcomb and McPherson 1994) as have other categories of semantically laden events (e.g., environmental sounds—Van Petten and Rheinfelder 1995). The general consensus is that the N400 reflects a process of semantic access via lexical or other kinds of representations. In language studies, the N400 thus appears to be sensitive to lexico-semantic processing.

With this background we now return to the issue of lexical neighborhood density. In one study in English monolinguals (Holcomb et al. 2002) words with a large

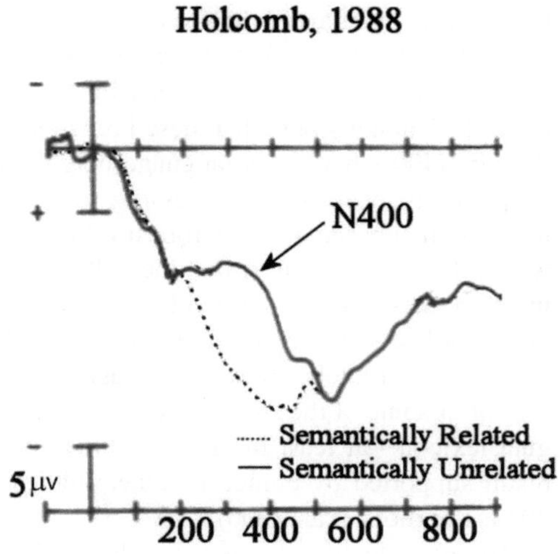

Fig. 7.1 ERPs from a right parietal site in a semantic priming lexical decision task

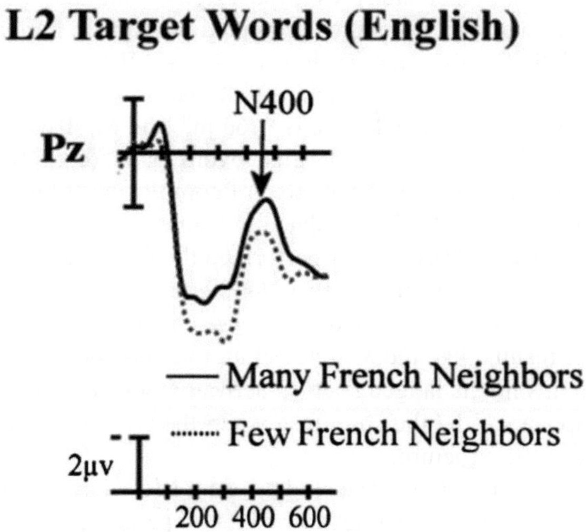

Fig. 7.2 N400 effects of neighborhood density in L1 (English) while processing L2 (French) words. Larger cross-language neighborhoods associated with greater N400 amplitude

number of orthographic neighbors (e.g., the word "time" has many words that are orthographically similar including "dime," "lime," "tine," "tame," etc.) were shown to generate a larger N400 component than words with a smaller number of neighbors (e.g., "yacht" which has no neighbors—see Fig. 7.2) suggesting that the lexical representations of all items in an orthographic neighborhood are partially activated when a particular word is encountered.

Midgley et al. (2008) exploited this finding and investigated the influence of cross-language neighbors on the N400 in an ERP study with French–English bilinguals. As in Holcomb et al. (2002) the number of neighbors (many versus few) in the target language modulated the N400 (larger N400 for words with more neighbors in the target language) both for L1 and for a weaker L2. More to the point Midgley et al. also found an effect of neighborhood size of the non-target language on the N400 component. Even when only French words were presented in a list the number of lexical neighbors that these French items had in English modulated the N400 (larger N400 when cross-language neighborhood size was higher—see Fig. 7.2). Surprisingly, even when processing in the stronger L1, the neighborhood size of items in the unseen and less robust L2 modulated the amplitude of the N400 to L1 words. Because we know that the N400 reflects processing at the lexical-semantic interface these ERP data provide more compelling evidence for an account of bilingual word processing whereby words in a bilingual speaker's two languages are accessed in parallel even when the task at hand does not require it. These data therefore offer some of the strongest evidence to date of an integrated nonselective bilingual lexicon—at least in languages that share a common orthographic system, a result supported by numerous other studies (Midgley et al. 2011; Ng and Wicha 2013; Kerkhofs et al. 2006).

The nonselective access view also finds support in a study by Thierry and Wu (2007). In their study, Chinese–English bilinguals performed a semantic relatedness task with pairs of words presented in English. Participants were unaware that 50% of the English pairs such as POST–MAILBOX consisted of English words whose Chinese translation contained a repetition of a Chinese character (e.g., the translation of POST in Chinese is "you jü," and MAILBOX translated in Chinese is "you xiang"), while other English word pairs (e.g., BATH–SHOWER) translated to Chinese words that consisted of completely different (non-repeated) Chinese characters (e.g., BATH translated in Chinese is "xi zao," while SHOWER is in Chinese "ling yu"). Participants performed the task in a purely English context. The finding of interest was that a larger N400 effect was found for English word pairs for which there was a repetition of a Chinese character in the Chinese translation. This pattern of results, like the Midgley et al. findings seems most consistent with the idea that items in one language cross-activate items in a bilingual's other language. And this appears to be the case even when the two languages do not share a similar orthographic structure.

Together the above studies using ERPs have definitively shown that across multiple languages and levels of L2 competence, there is little evidence in favor of the notion of separate (i.e., language-selective) lexical systems in bilingual language users. Moreover, the results from these studies demonstrate their power to take us beyond the evidence of studies using offline measures of processing.

7.2.2 Language Control

The above review strongly suggests that bilingual individuals do not store or access words in their two languages from different autonomous lexical systems. While this might be seen as making the task of laying down new lexical representations in L2 somewhat easier, such an arrangement presents the bilingual language user with another problem—how to keep their two languages straight during processing. In other words, with a simple integrated lexicon it would seem as though a language user would frequently be making mistakes in both production and comprehension trying to decide which language to access. What is clear is that developing a mechanism(s) to suppress the unused language during normal language use is imperative to clear bilingual communication. A related issue, but almost opposite in terms of functional outcome, is the apparent fluidity with which many bilinguals *switch* (more accurately *code-switch*) between their two languages in certain language contexts. Both of these observations about bilingual language use may be considered under the common heading of *Language Control*. Language control has been extensively studied and has come to dominate much of the psycholinguistic and neuro-cognitive literature on bilingualism. There is an interesting paradox surrounding bilingual language control. While the cognitive demands of language control have been observed to negatively impact bilingual processing (Gollan et al. 2005) recently, it has also been argued that exerting such control might also have certain "cognitive" advantages (for a review, see Bialystok and Craik 2010).

While the location of control mechanisms in the brain using functional imaging techniques (primarily fMRI) has tended to dominate this literature, there have been a number of electrophysiological studies looking at language control from a processing standpoint. Much of this literature has looked at language production (e.g., Jackson et al. 2001, 2004; Christoffels et al. 2007; Strijkers et al. 2010) although a few studies have examined comprehension as well (Chauncey et al. 2008, 2011). Here we will concentrate on the studies of comprehension. In a study by Alvarez et al. (2003) native-English speakers enrolled in beginning and intermediate university Spanish courses participated in a mixed language semantic categorization task in which critical words were presented in English (L1) and Spanish (L2). ERPs were recorded to all words. The important finding here was that there was a greater negativity in the ERP waveforms when the word in the directly preceding trial was from the other language compared to when it was from the same language. Within the time frame of the N400, this language switch effect arose only when the target word was Spanish and the preceding word was English (i.e., an L1–L2 switch), however, there was a later effect (in the same direction) for L2–L1 switches (see Fig. 7.3). The authors speculated that there is a general language-switch effect on ERPs that takes longer to develop in the L2–L1 direction. This effect could be related to the inhibitory control over lexical activation proposed in models such as the Bilingual Interactive Activation (BIA) model (Van Heuven et al. 1998). According to this model, presentation of a word in one language leads to the global inhibition of all words in the other language. Accordingly, this inhibition will make it harder to process words following a language switch, as evident in behavioral studies (Grainger and Beauvillain 1987; Thomas and Allport 2000). The inhibition of word-form representations will make it harder to recover the associated meaning, thus causing an increase in N400 amplitude. The critical aspect of the Alvarez et al. result was that a language-switch effect was observed in ERP measures when participants did not have to make language-selective responses to critical word stimuli. This suggests that language switches are affecting processing related to the recovery

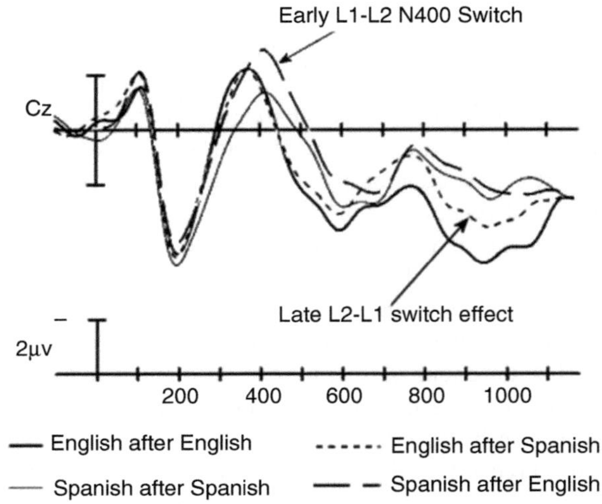

Fig. 7.3 Language switching effects from both directions, L1 to L2 and L2 to L1 (Adapted from Alvarez et al. (2003))

of meaning from form and are not the result of task-specific decision schemas that operate separately on each language (Green 1998). Most important, these language switch effects should be clearly distinguished from switch costs that are observed when a given language must be selected in order to perform the task.

In Chauncey et al. (2008) two experiments were used to test language switching effects with bilingual participants, but now in a paradigm where overt strategic influences could be minimized. Motivated by the findings of Alvarez et al. (2003) a priming paradigm with masked primes was used. One problem with the conclusions from Alvarez et al, that the presence of the other (stronger) language produced a cost as indicated by a larger N400 on switch trials, is that participants were aware of the switches and might have used some type of strategic processing that enhanced the effect. In Chauncey et al. participants had to monitor target words for animal names, and ERPs were recorded to critical (non-animal) words in L1 and L2 primed by unrelated words from the same or the other language—but presented too quickly to result in a differential strategy. Both experiments revealed language priming (switching) effects that depended on target language. For target words in L1, most of the language switch effect appeared in the N400 ERP component, with L2 primes generating a more negative-going wave than L1 primes. For L2 target words, on the other hand, the effects of a language switch appeared mainly in an earlier ERP component (the N250) peaking at approximately 250 ms post-target onset, and showing greater negativity following an L1 prime than an L2 prime (see Fig. 7.4). This is the first evidence for fast-acting language-switching effects occurring in the absence of overt (conscious) task switching and suggests that some of the top-down influences in models such as the BIA can occur without conscious awareness.

In a second study Chauncey et al. (2011) found that the effects of switching language across prime and target differed as a function of the direction of the switch and prime duration. Effects tended to be stronger with 100 ms prime durations than with 50 ms durations, and the expected pattern of greater negativity in the switch

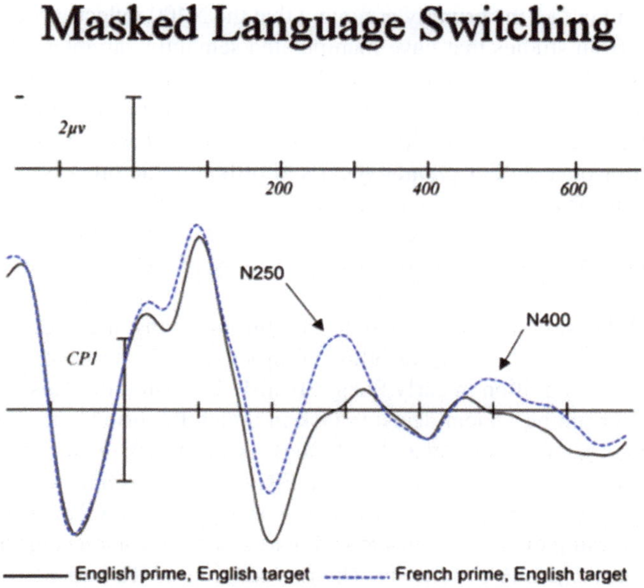

Fig. 7.4 Masked language switch effects on L2 (English) items; effects predominately on the N250 component (Adapted from Chauncey et al. 2008)

condition appeared earlier when primes were in L1 and targets in L2 than vice versa. The authors also examined whether these language-switching effects could be due to differences in the subjective frequency of words in a bilingual's two languages, by testing a frequency-switching manipulation within L1. Effects of frequency switching (e.g., low frequency prime, high frequency target) were evident in the ERP waveforms, but the pattern did not resemble the language switching effects.

Taken together the results suggest that different mechanisms are at play, and, in the case of language switches, probably involve a mechanism such as the language nodes in the BIA model.

7.2.3 Language Interactivity

Up to this point we have mainly been looking at the electrophysiological evidence for nonselective lexical representation and interactivity of lexical processing in two languages. In the following studies we will be examining more directly and more extensively the interactivity of processing across levels of representation which is afforded when priming techniques are employed.

Priming has a rich history in experimental psychology as a technique that reveals interactions between levels of processing. In the case of semantic priming (shown above for the N400) it is clear that the interaction is mainly at the level of semantic processing and in fact the primary argument that the N400 reflects processing at this level comes from studies that have manipulated semantic variables. However, it is also possible to look at interactions at earlier points in processing, prior to semantic activation, by manipulating other relationships between prime and target stimuli. For example, by using primes and targets that rhyme or share letters in common, it is possible to isolate earlier phonological and orthographic processes. By combining ERPs with priming it is then possible to examine the time course of various processes involved in word recognition. And by including a bilingual manipulation as well, one can use this approach to study a variety of second language processing questions.

Kotz (2001) was one of the first to combine priming methodology with the recording of ERPs in the study of bilingual processing. In her study she set out to examine word recognition in early fluent Spanish–English bilinguals using a word-pair semantic priming lexical decision task in which the prime–target relationship could be either categorically related (cat–dog), associatively related (honey-bee), or unrelated (truck-hand). Reaction times and ERPs were measured while participants viewed blocks of either Spanish or English words. Whereas RT measures reflected associative priming effects, ERPs showed both associative and categorical priming effects in both language conditions. The dissociation of RT and ERP effects suggests that the two measures might tap into different underlying processes during semantic priming. Importantly, both RT and ERP measures revealed symmetrical priming in L1 and L2 indicating that semantic processing in early fluent bilinguals is equivalent for L1 and L2.

In the Alvarez et al. (2003) study mentioned above there was also a priming manipulation. Recall that their study used native-English speakers enrolled in beginning and intermediate university Spanish courses. Their participants engaged in a mixed language semantic categorization task in which critical words were presented in English (L1) and Spanish (L2); repetitions of these words (within- and between languages) were also presented on subsequent trials (i.e., immediate repetition). Repetition priming is thought to represent a stronger priming manipulation than semantic priming as it taps processing at both the semantic and form levels of processing and in this sense might be expected to be a more sensitive paradigm for assessing bilingual effects. ERPs were recorded to all items allowing for comparisons of the N400 component to repetitions within- and between languages as well as to words presented for the first time. Two important findings were observed in their study during relatively early stages of acquiring a second language. First, in the typical N400 window (300–500 ms), between-language repetition (translation) produced a smaller reduction in N400 amplitude than did within-language repetition (see Fig. 7.5). And second, the time course of between-language repetition effects tended to be more extended in time and differed as a function of language with L2–L1 repetitions producing larger priming effects early (during the typical N400 window) and L1–L2 repetitions producing larger priming effects later (during windows after the typical N400). So, unlike Kotz (2001) who, testing proficient bilinguals, demonstrated symmetrical priming in L1 and L2, Alvarez et al. showed asymmetries in priming in adult bilinguals still acquiring their L2. That variation in

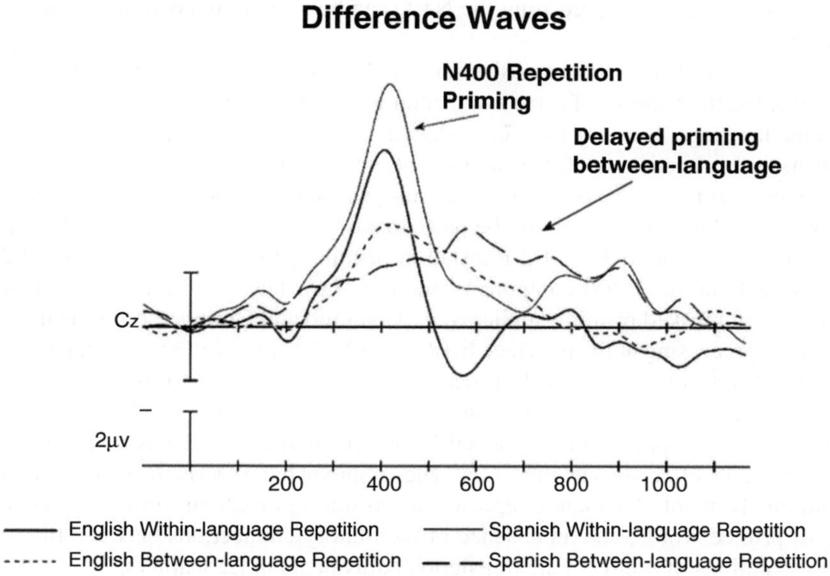

Fig. 7.5 Repetition priming effects, within and between languages (Adapted from Alvarez et al. 2003)

the N400 amplitude has been found to be correlated with efficient L2 use supports its use as a sensitive measure of language proficiency.

Building on these earlier results, Geyer et al. (2011) examined 20 native-Russian speakers who were highly proficient in English. Participants engaged in a mixed-language lexical decision task in which critical words were presented in Russian (L1) and English (L2) and repetitions of these words (within and between languages) were presented on subsequent trials. ERPs were recorded to all items allowing for comparisons of repetition effects within and between (translation) languages. The results revealed a symmetrical pattern of within-language repetition and between-language translation ERP priming effects, which in conjunction with Alvarez et al. (2003) and Kotz (2001) supports the hypothesis that L2 proficiency level rather than age or order of language acquisition is responsible for the observed patterns of translation priming.

Recall that at the beginning of this section we argued that using ERPs in conjunction with priming had the advantage of revealing multiple sources of priming effects. The above studies all showed effects either only on the N400 component or on the N400 and later components in bilinguals. So one question is why aren't earlier pre-semantic effects showing up at least in the case of repetition priming where differences across both pre and post-lexical processing might be expected? One possible answer is that typical priming studies aren't able to see earlier effects. In fact Holcomb and Grainger (2006) argued just this in their pioneering study combining masked priming and ERP recordings. In their experiments they showed that at least three ERP components preceding the N400 can be seen in this paradigm which we explore below.

Midgley et al. (2009a) used the same ERP and masked priming paradigm to examine the time course of form and meaning activation during word recognition in second language learners in a cross-language (translation) priming experiment. Targets were repetitions of, translations of, or were unrelated to the immediately preceding prime. In Experiment 1 all targets were in the participants' L2. In Experiment 2 all targets were in the participants' L1. In Experiment 1 both within-language repetition and L1–L2 translation priming produced effects on the N250 component and the N400 component (see Fig. 7.6). The N250 a component, seen primarily in masked priming studies (e.g., Holcomb and Grainger 2006), is thought to reflect processing at the interface between orthographic form and lexical processing (the N400 reflects primarily processing at the later word-form/meaning interface). In Experiment 2 only within-language repetition produced N250 effects, while both types of priming produced N400 effects (i.e., there was no evidence of L2–L1 N250 effects—see Fig. 7.7). These somewhat surprising results suggest rapid involvement of semantic representations during on-going form-level processing of printed words, and an absence of facilitatory connections between the form representations of non-cognate translation equivalents in L2 learners.

These findings raise questions about translation priming across different orthographic scripts. To this end, Hoshino et al. (2010) examined the time course of cross-script translation priming and repetition priming in two different scripts using

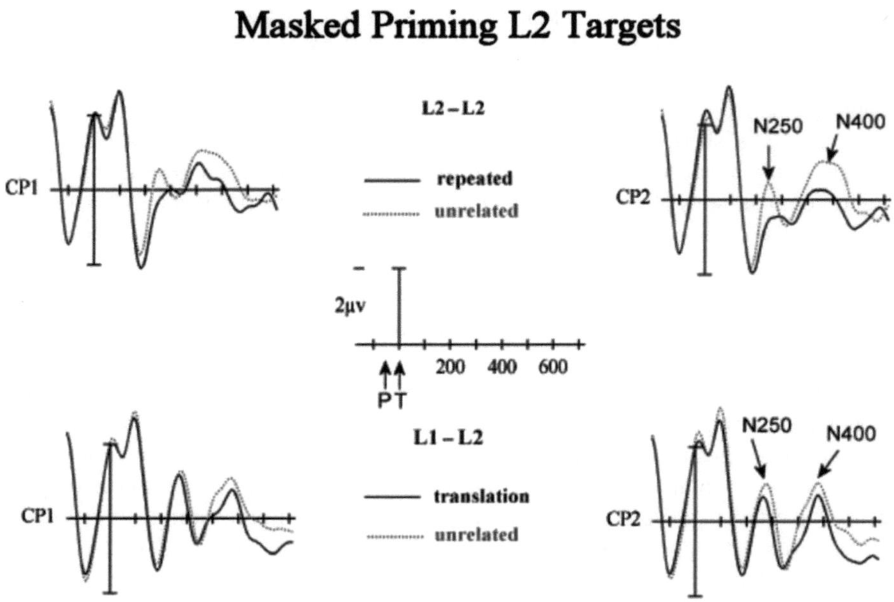

Fig. 7.6 Masked repetition priming effects in L2 and L1 to L2. (Adapted from Midgley et al. (2009a, b))

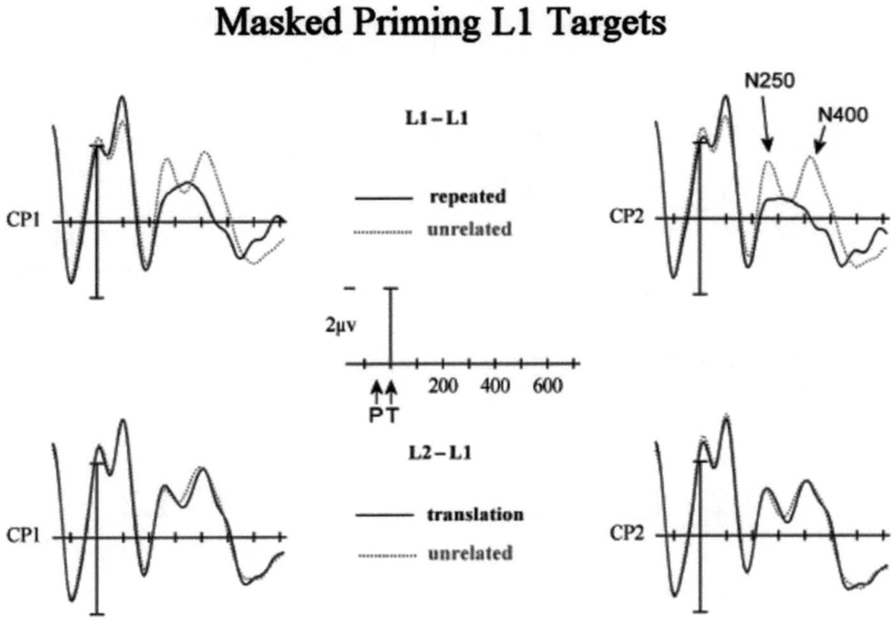

Fig. 7.7 Masked repetition priming effects in L1 and L2 to L1. (Adapted from Midgley et al. (2009a))

a combination of the masked priming paradigm with the recording of ERPs. Japanese–English bilinguals performed a semantic categorization task in their second language (English) and in their first language (Japanese). Targets were preceded by a visually presented related (translation equivalent/repeated) or unrelated prime. The results showed that the amplitudes of the N250 and N400 ERP components were significantly modulated for L2–L2 repetition priming, L1–L2 translation priming, and L1–L1 repetition priming, but not for L2–L1 translation priming (see Fig. 7.8). There was also evidence for priming effects in an earlier 100–200 ms time window for L1–L1 repetition priming and L1–L2 translation priming. They argued that a change in script across primes and targets provides optimal conditions for prime word processing, hence generating very fast-acting translation priming effects when primes are in L1.

In a follow-up to these earlier findings, Schoonbaert et al. (2011) had English–French bilinguals perform a lexical decision task while ERPs were measured to L2 targets, preceded by L1 translation primes versus L1 unrelated primes (Experiment 1a) and vice versa (Experiment 1b). The big difference between this and the Midgley et al. study was that the prime–target stimulus onset asynchrony was 120 ms (versus 70 ms in Midgley et al.). As in Midgley et al. significant masked translation priming was observed in the L1–L2 direction on both the N250 and the N400 (see Fig. 7.9 right). However, unlike Midgley et al., there was also a strong and early N250 L2 to L1 priming effect (see Fig. 7.9 left). This result indicates that L2 primes can influence L1 processing even in the rapid masked priming paradigm if the prime words

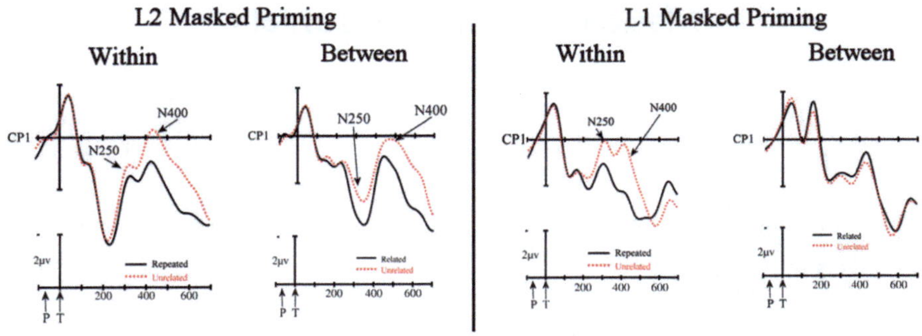

Fig. 7.8 Masked repetition priming effects across scripts within and between languages (Adapted from Hoshino et al. (2010))

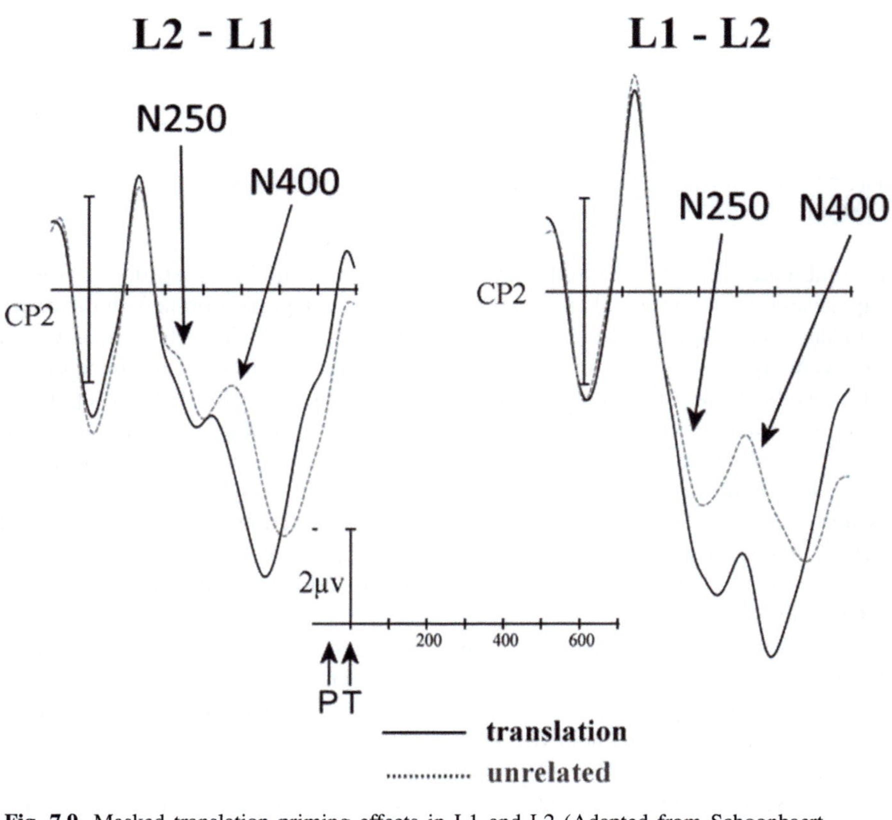

Fig. 7.9 Masked translation priming effects in L1 and L2 (Adapted from Schoonbaert et al. 2011)

are given a bit more time to be processed (30 ms in this case) prior to onset of the L1 target and suggests that one of the big differences between L1 and L2 processing is in the sluggishness of pre-lexical mechanisms in a less competent language. Together the above studies reflect the richness of the priming paradigm for uncovering interactivity between form and meaning at different levels of a bilingual's two languages and in different time windows evident in behavioral and ERP data.

7.2.4 Language Effects

While all of the studies reviewed up to this point have involved some type of psycholinguistic manipulation to modulate differences in L1 and L2 processing, one of the big advantages of ERPs is that it is also possible to simply look for differences in the ERP signal between L1 and L2 words during language processing. This is because the pattern of ERP components to L1 words is quite well understood and so departures from this pattern can be meaningfully interpreted as reflecting differences in word processing in another language—what we will call here ERP language effects.

In Liu and Perfetti (2003) Chinese bilinguals performed a delayed naming task, reading both Chinese characters and English words, while ERPs were recorded. Using Principle Component Analysis (PCA) they reported effects from the onset of the stimulus that suggested a temporal unfolding of graphic, phonological, and semantic processing that depended on both language and word frequency. At 150 ms, Chinese produced an earlier and higher amplitude shift (N150) than English. At 250 ms, frequency effects were significant for both Chinese and English, but at 450 ms, only the English frequency effect was reliable.

Midgley et al. (2009b) examined language effects in second language learners. In three experiments participants monitored a stream of words for occasional probes from one semantic category and ERPs were recorded to non-probe critical items. In Experiment 1 L1 English participants who were university learners of French saw two lists of words blocked by language, one in French and one in English. We observed a large effect of language that mostly affected the amplitudes of the N400 component, but starting as early as 150 ms post-stimulus onset (Fig. 7.10 left). A similar pattern was found in Experiment 2 with L1 French and L2 English, showing that the effect is due to language dominance and not language per se (Fig. 7.10 middle).

Experiment 3 found that proficient French/English bilinguals exhibited a different pattern of language effects suggesting that these effects are modulated by proficiency (Fig. 7.10 right). These results lend further support to the hypothesis that word recognition during the early phases of L2 acquisition in late learners of L2 involves a specific set of mechanisms, compared with recognition of L1 words.

Results from speakers of more than two languages have yielded different results. Aparicio et al. (2012) recorded ERPs during the visual presentation of words in the three languages of French–English–Spanish trilinguals while performing a semantic

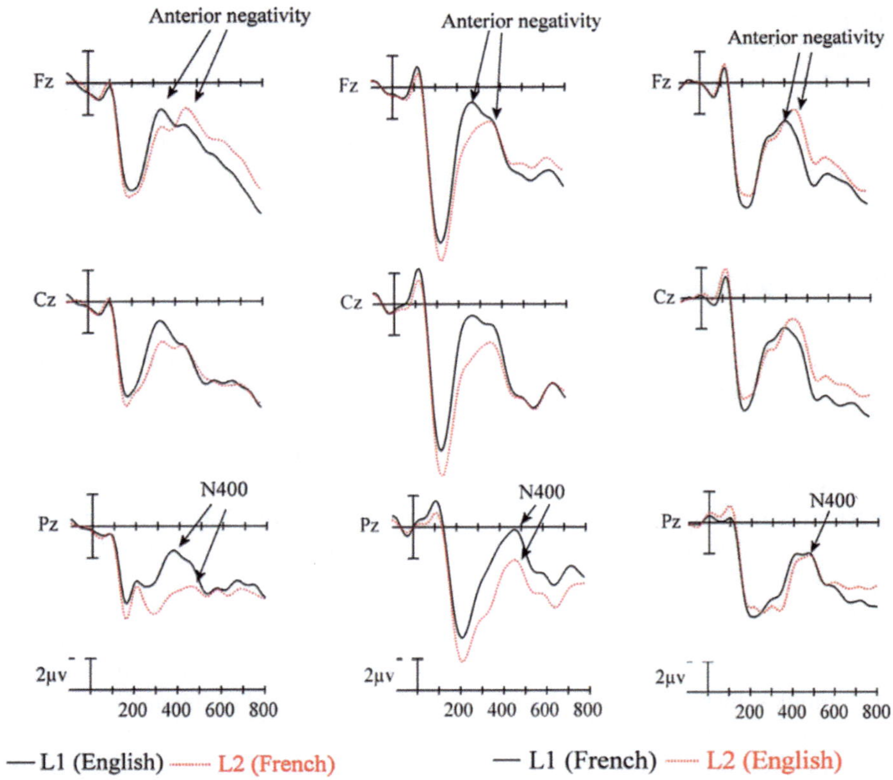

Fig. 7.10 Language effects in L1 and L2 (Adapted from Midgley et al. (2009a, b)

categorization task. Words in L1 generated earlier N400 peak amplitudes than both L2 and L3 words, which peaked together. On the other hand, L2 and L3 words did differ significantly in terms of N400 amplitude, with L3 words generating greater mean amplitudes compared with L2 words. They interpreted the effects of peak N400 latency as reflecting the special status of the L1 relative to later acquired languages, rather than proficiency in that language per se. On the other hand, the mean amplitude difference between L2 and L3 is thought to reflect different levels of subjective frequency in these two languages.

Studies with monolingual adults have identified successive stages occurring in different brain regions for processing single written words. Leonard et al. (2010) combined MEG and MRI to compare L1 and L2 in bilingual adults. L1 words in a size judgment task evoked a typical left-lateralized sequence of activity first in ventral occipitotemporal cortex (VOT, visual word-form encoding) and then ventral frontotemporal regions (lexico-semantic processing). Compared to L1, words in L2 activated right VOT more strongly from ~135 ms; this activation was attenuated when words became highly familiar with repetition. At ~400 ms, L2 responses were generally later than L1, more bilateral, and included the same lateral occipitotemporal areas as were activated by pictures. The authors propose that acquiring a new language involves the recruitment of right hemisphere and posterior visual areas that are not necessary once fluency is achieved. These findings also allow us to interpret differences in word processing attributable to language dominance effects.

7.2.5 Language Learning

In both proficient bilinguals (i.e., those nearly equally competent in both languages) or university learners in more advanced L2 language courses, it is almost impossible to infer backwards in time to the point where a new L2 first starts to take hold in the brain. In this section we review a few studies that attempted to examine such changes.

McLaughlin et al. (2004) used ERPs in a study of second-language (L2) learning. They pointed out that adult learning is often claimed to be slow and laborious compared to native language (L1) acquisition, but little is known about the rate of L2 word learning. Here we report that adult second-language learners' brain activity, as measured by ERPs, discriminated between L2 words and L2 "pseudowords" (word-like letter strings) after just 14 h of classroom instruction. This occurred even while the learners performed at chance levels when making overt L2 word/pseudoword judgments, indicating that the early acquisition of some aspects of a new language may be overlooked by current behavioral assessments.

Liu, Perfetti, and Wang (2006) used ERPs in learners of Chinese (at the end of their first and second terms of Chinese class at an American university) to assess the learning of word-form, pronunciation, and meaning in an unfamiliar writing system. The participants were required to recognize a target Chinese character or English word while ERPs were recorded. They named filler targets indicated by a signal 1000 ms after the onset of the stimuli. The orthographic processing of characters and words was extracted as a 200 ms component by Principle Component Analysis. The semantic processing was extracted as a 400 ms component (N400). The 200 ms PCA component was negative at occipital electrodes (N200) and positive at frontal electrodes (P200). There was sensitivity to visual analysis and lexical access, respectively. ERP results showed that the visual analysis of Chinese was more difficult than English during the first term, but not the second term. The lexical access was more difficult and the semantic processing was slower for Chinese than English at both terms. Faster lexical access was obtained for familiar characters during the first term, but not the second term. The separation of visual analysis and lexical access during the second term indicates a threshold style processing of Chinese characters for the learners with moderate reading proficiency.

These effects draw our attention to the issue of whether there is an absolute critical period for acquiring language (Ojima et al. 2005). One approach to address this issue is to compare the processes of second language (L2) learning after childhood and those of first language (L1) learning during childhood. To study the cortical process of post-childhood L2 learning, Ojima and colleagues compared event-related brain potentials recorded from two groups of adult Japanese speakers who attained either high or intermediate proficiency in English after childhood (J-High and J-Low), and adult native English speakers (ENG). Semantic anomalies embedded in English sentences evoked a clear N400 component in all three groups, with only the time course of the brain activation varying among the groups. Syntactic violations elicited a left-lateralized negativity similar to the left anterior negativity in ENG and J-High, but not in J-Low. In ENG, a P600 component was additionally found. These results suggest that semantic processing is robust from early on in L2 learning, whereas the development of syntactic processing is more dependent on proficiency as evidenced by the lack of the left-lateralized negativity in J-Low. Because early maturation and stability of semantic processing as opposed to syntactic processing are also a feature of L1 processing, post-childhood L2 learning may be governed by the same brain properties as those which govern childhood L1 learning. It is possible that these processes are qualitatively similar in many respects, with only restricted domains of language processing being subject to absolute critical period effects.

Stein et al. (2006) used ERPs to trace changes in brain activity related to progress in second language learning. Twelve English-speaking exchange students learning Swiss German in Switzerland were recruited. ERPs to visually presented single words from the subjects' native language (English), second language (German), and an unknown language (Romansh) were measured before (first session: day 1) and after (second session: day 2) 5 months of intense German language learning. When comparing ERPs to English words, they found differences between 472 and 644 ms across days. In ERPs to Romansh words, no differences were observed.

Together these results seem to reflect plasticity in the neuronal networks underlying second language acquisition and indicate that with a higher level of second language proficiency, second language word processing is faster. Yum et al. (2014) examined the very initial phases of orthographic and semantic acquisition in monolingual English speakers learning Chinese words under controlled laboratory conditions. They found that only the fast learners showed a left-lateralized increase in the N170 amplitude with training (see Fig. 7.11). Furthermore, only the fast learners showed an increased N400 amplitude with training, with a distinct anterior distribution. Slow learners, on the other hand, showed a positive posterior effect, with increasingly positive-going waveforms in occipital sites as training progressed (see Fig. 7.12).

7.3 Concluding Remarks

In this chapter we review electrophysiological studies of word processing in bilingualism. We see great promise in these techniques for uncovering important neurocognitive underpinnings of bilingual language processing—in particular the time course of the subcomponents of word processing during language comprehension and the interaction between these components. Combining ERP with priming

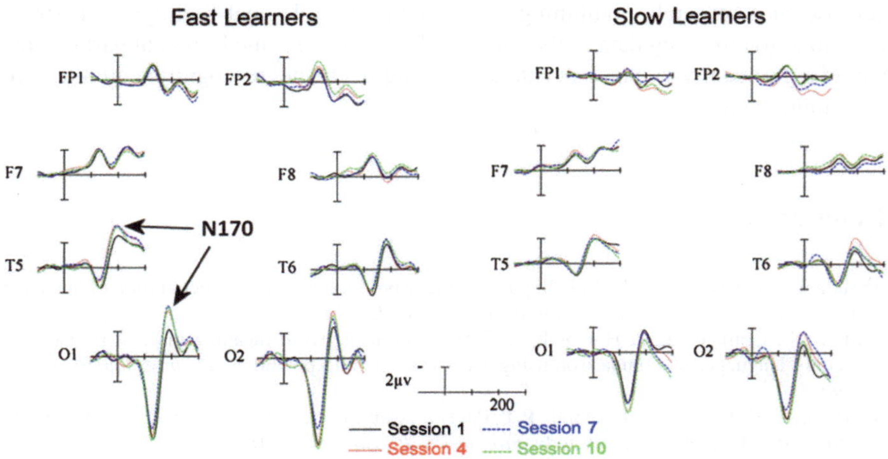

Fig. 7.11 N170 effects in fast vs. slow L2 learners (Adapted from Yum et al. (2014))

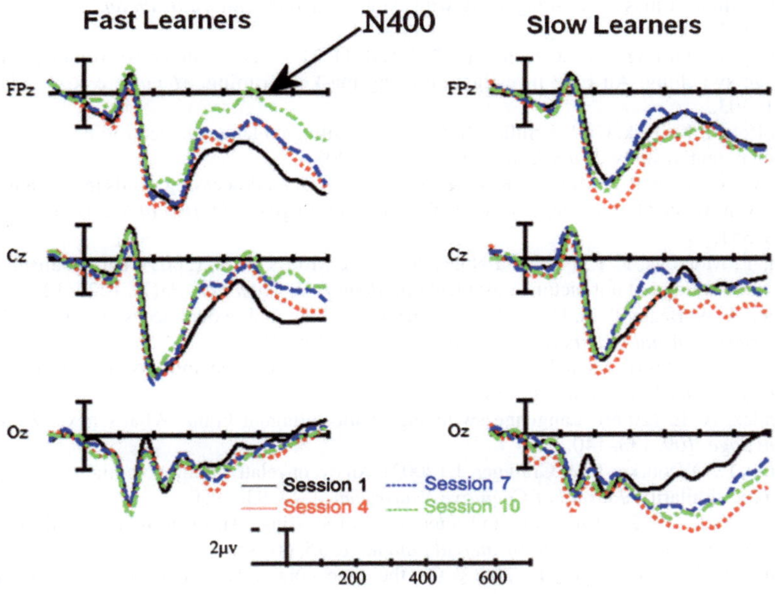

Fig. 7.12 Developing N400 effects during L2 learning in fast vs. slow learners. (Adapted from Yum et al. (2014))

paradigms has proven to be a rich source of information about processing in the bilingual brain. The impact of language difference, selectivity, control, and interactions on bilingual processing highlights just some of the complexities of this topic and the importance of time-sensitive techniques like EEG and MEG in examining these variables. With the growing emphasis on multimodal integration in cognitive neuroscience research, combining these electrophysiological findings with structural/function imaging data in the same studies with the same bilingual participants would significantly enhance our understanding of the neuro-cognitive architecture of bilingual brain.

References

Abutalebi, J., & Green, D. (2007). Bilingual language production: The neurocognition of language representation and control. *Journal of Neurolinguistics, 20*, 242–275.

Alvarez, R., Grainger, J., & Holcomb, P. (2003). Accessing word meaning in primary and secondary languages: A comparison using event-related brain potentials. *Brain & Language, 87*, 290–304.

Aparicio, X., Midgley, K. J., Holcomb, P. J., Pu, H., Lavaur, J.-M., & Grainger, J. (2012). Language effects in trilinguals: An ERP study. *Frontiers in Psychology, 3*, 402.

Bialystok, E., & Craik, F. I. (2010). Cognitive and linguistic processing in the bilingual mind. *Current Directions in Psychological Science, 19*, 19–23.

Chauncey, K., Grainger, J., & Holcomb, P. J. (2008). Code-switching effects in bilingual word recognition: A masked priming study with event-related potentials. *Brain and Language, 105*, 161–174.

Chauncey, K., Grainger, J., & Holcomb, P. J. (2011). The role of subjective frequency in language switching: An ERP investigation using masked priming. *Memory & Cognition, 39*, 291–303.

Christoffels, I. K., Firk, C., & Schiller, N. O. (2007). Bilingual language control: An event-related brain potential study. *Brain Research, 1147*, 192–208.

Geyer, A., Holcomb, P. J., Midgley, K. J., & Grainger, J. (2011). Processing words in two languages: An event-related brain potential study of proficient bilinguals. *Journal of Neurolinguistics, 24*, 338–351.

Gollan, T., Montoya, R. I., Fennema-Notestine, C., & Morris, S. K. (2005). Bilingualism affects picture naming but not picture classification. *Memory & Cognition, 33*(7), 1220–1234.

Grainger, J., & Beauvillain, C. (1987). Language blocking and lexical access in bilinguals. *The Quarterly Journal of Experimental Psychology, 39*, 295–319.

Green, D. W. (1998). Mental control of the bilingual lexico-semantic system. *Bilingualism: Language and Cognition, 1*, 67–81.

Hernandez, A. E. (2009). Language switching in the bilingual brain: What's next? *Brain and Language, 109*, 133–140.

Holcomb, P., O'Rourke, T., & Grainger, J. (2002). An event-related brain potential study of orthographic similarity. *Journal of Cognitive Neuroscience, 14*, 938–950.

Holcomb, P. J. (1988). Automatic and attentional processing: An event-related brain potential analysis of semantic priming. *Brain and Language, 35*, 66–85.

Holcomb, P. J., & Grainger, J. (2006). On the time course of visual word recognition: An event-related potential investigation using masked repetition priming. *Journal of Cognitive Neuroscience, 18*, 1631–1643.

Holcomb, P. J., & McPherson, W. B. (1994). Event-related brain potentials reflect semantic priming in an object decision task. *Brain and Cognition, 24*, 259–276.

Hoshino, N., Midgley, K. J., Holcomb, P. J., & Grainger, J. (2010). An ERP investigation of masked cross-script translation priming. *Brain Research, 1344*, 159–172.

Jackson, G. M., Swainson, R., Cunnington, R., & Jackson, S. R. (2001). ERP correlates of executive control during repeated language switching. *Bilingualism: Language and Cognition, 4*, 169–178.

Jackson, G. M., Swainson, R., Mullin, A., Cunnington, R., & Jackson, S. R. (2004). ERP correlates of a receptive language-switching task. *Quarterly Journal of Experimental Psychology Section A, 57*, 223–240.

Kerkhofs, R., Dijkstra, T., Chwilla, D. J., & De Bruijn, E. R. (2006). Testing a model for bilingual semantic priming with interlingual homographs: RT and N400 effects. *Brain Research, 1068*, 170–183.

Kotz, S. A. (2001). Neurolinguistic evidence for bilingual language representation: A comparison of reaction times and event-related brain potentials. *Bilingualism: Language and Cognition, 4*, 143–154.

Kotz, S. A. (2009). A critical review of ERP and fMRI evidence on L2 syntactic processing. *Brain and Language, 109*, 68–74.

Kutas, M., & Hillyard, S. A. (1980). Reading senseless sentences: Brain potentials reflect semantic incongruity. *Science, 207*, 203–205.

Leonard, M. K., Brown, T. T., Travis, K. E., Gharapetian, L., Hagler, D. J., Dale, A. M., et al. (2010). Spatiotemporal dynamics of bilingual word processing. *Neuroimage, 49*, 3286–3294.

Liu, Y., & Perfetti, C. A. (2003). The time course of brain activity in reading English and Chinese: An ERP study of Chinese bilinguals. *Human Brain Mapping, 18*, 167–175.

Liu, Y., Perfetti, C. A., & Wang, M. (2006). Visual analysis and lexical access of Chinese characters by Chinese as second language readers. *Language and Linguistics, 7*, 637–657.

Luck, S. J. (2014). *An introduction to the event-related potential technique*. Cambridge, MA: MIT Press.

McLaughlin, J., Osterhout, L., & Kim, A. (2004). Neural correlates of second-language word learning: Minimal instruction produces rapid change. *Nature Neuroscience, 7*, 703–704.

Midgley, K. J., Holcomb, P. J., & Grainger, J. (2009a). Masked repetition and translation priming in second language learners: A window on the time-course of form and meaning activation using ERPs. *Psychophysiology, 46*, 551–565.

Midgley, K. J., Holcomb, P. J., & Grainger, J. (2009b). Language effects in second language learners and proficient bilinguals investigated with event-related potentials. *Journal of Neurolinguistics, 22*, 281–300.

Midgley, K. J., Holcomb, P. J., & Grainger, J. (2011). Effects of cognate status on word comprehension in second language learners: An ERP investigation. *Journal of Cognitive Neuroscience, 23*, 1634–1647.

Midgley, K. J., Holcomb, P. J., Walter, J., & Grainger, J. (2008). An electrophysiological investigation of cross-language effects of orthographic neighborhood. *Brain Research, 1246*, 123–135.

Ng, S., & Wicha, N. Y. (2013). Meaning first: A case for language-independent access to word meaning in the bilingual brain. *Neuropsychologia, 51*, 850–863.

Nunez, P. L., & Srinivasan, R. (2006). *Electric fields of the brain: The neurophysics of EEG*. New York: Oxford University Press.

Ojima, S., Nakata, H., & Kakigi, R. (2005). An ERP study of second language learning after childhood: Effects of proficiency. *Journal of Cognitive Neuroscience, 17*, 1212–1228.

Schoonbaert, S., Holcomb, P. J., Grainger, J., & Hartsuiker, R. J. (2011). Testing asymmetries in noncognate translation priming: Evidence from RTs and ERPs. Psychophysiology, 48, 74–81.

Stein, M., Dierks, T., Brandeis, D., Wirth, M., Strik, W., & König, T. (2006). Plasticity in the adult language system: A longitudinal electrophysiological study on second language learning. *Neuroimage, 33*, 774–783.

Strijkers, K., Costa, A., & Thierry, G. (2010). Tracking lexical access in speech production: Electrophysiological correlates of word frequency and cognate effects. *Cerebral Cortex, 20*, bhp153.

Thierry, G., & Wu, Y. J. (2007). Brain potentials reveal unconscious translation during foreign-language comprehension. *Proceedings of the National Academy of Sciences, 104*, 12530–12535.

Thomas, M. S., & Allport, A. (2000). Language switching costs in bilingual visual word recognition. *Journal of Memory and Language, 43*, 44–66.

van Heuven, W. J., & Dijkstra, T. (2010). Language comprehension in the bilingual brain: fMRI and ERP support for psycholinguistic models. *Brain Research Reviews, 64*, 104–122.

Van Heuven, W. J., Dijkstra, T., & Grainger, J. (1998). Orthographic neighborhood effects in bilingual word recognition. *Journal of Memory and Language, 39*, 458–483.

Van Petten, C., & Kutas, M. (1991). Electrophysiological evidence for the flexibility of lexical processing. *Advances in Psychology, 77*, 129–174.

Van Petten, C., & Rheinfelder, H. (1995). Conceptual relationships between spoken words and environmental sounds: Event-related brain potential measures. *Neuropsychologia, 33*, 485–508.

Yum, Y. N., Midgley, K. J., Holcomb, P. J., & Grainger, J. (2014). An ERP study on initial second language vocabulary learning. *Psychophysiology, 51*, 364–373.

Part III
Language: Looking Back in Time and Space

Chapter 8
Causal Modeling: Methods and Their Application to Speech and Language

Baojuan Li, Seppo P. Ahlfors, Dimitris Pinotsis, Karl J. Friston, and Maria Mody

8.1 Introduction

Cognitive task performance relies upon distributed networks of interacting brain regions. These networks can be studied with neuroimaging techniques, such as functional magnetic resonance imaging (fMRI), electroencephalography (EEG),

B. Li
Athinoula A. Martinos Center for Biomedical Imaging, Charlestown, MA 02129, USA

Department of Radiology, Massachusetts General Hospital and Harvard Medical School, Boston, MA 02114, USA

School of Biomedical Engineering, Fourth Military Medical University, Xi'an, Shaanxi 710032, China
e-mail: libjuan@163.com

S.P. Ahlfors • M. Mody (✉)
Athinoula A. Martinos Center for Biomedical Imaging, Charlestown, MA 02129, USA

Department of Radiology, Massachusetts General Hospital and Harvard Medical School, Boston, MA 02114, USA
e-mail: seppo@nmr.mgh.harvard.edu; maria@nmr.mgh.harvard.edu

D. Pinotsis
The Picower Institute for Learning & Memory, Massachusetts Institute of Technology, Cambridge, MA 02139, USA

Department of Brain and Cognitive Sciences, Massachusetts Institute of Technology, Cambridge, MA 02139, USA

The Wellcome Trust Centre for Neuroimaging, University College London, Bloomsbury, London WC1N 3BG, UK
e-mail: pinotsis@mit.edu

K.J. Friston
The Wellcome Trust Centre for Neuroimaging, University College London, Bloomsbury, London WC1N 3BG, UK
e-mail: k.friston@ucl.ac.uk

M. Mody (ed.), *Neural Mechanisms of Language*, Innovations in Cognitive Neuroscience, DOI 10.1007/978-1-4939-7325-5_8

and magnetoencephalography (MEG). To investigate the interactions between the brain regions, two types of measures have been proposed: functional connectivity and effective connectivity (Friston 1994). *Functional connectivity* is a simple but useful measure to describe temporal correlations (i.e., statistical dependencies) between the activities of different brain regions or sources. Functional connectivity does not provide any explicit information about how distributed activity is caused and propagated within a network. However, characterizing directed information flow within a brain network is crucial for understanding both normal and impaired brain function. Directed information flow is highly relevant in the neuronal processing of speech and language as they rely on both bottom-up and top-down mechanisms. Currently, it is possible to make inferences about *directed functional connectivity* by appealing to temporal precedence. This approach is embodied in measures like Granger causality (GC) and transfer entropy (TE) that are based upon time series from different sources or recording channels. An alternative approach to characterizing directed coupling between brain regions starts with a forward or generative model of distributed processing and then tries to estimate the (coupling) parameters of that model. The resulting estimates of directed coupling are referred to as *effective connectivity*, which refers to the causal influence one brain region or system exerts over another (Friston 1994). Effective connectivity analysis, such as dynamic causal modeling (DCM), aims at characterizing causal interactions between distributed brain regions involved in a cognitive process and—crucially—enables one to test hypotheses about condition or context-sensitive changes in connectivity that generally elude functional connectivity approaches. This allows one to test different hypotheses or network architectures in terms of how well they explain experimental data—and quantify condition or cohort-specific effects in terms of changes in effective connectivity.

In this chapter, we first review select studies that have used causal modeling to investigate the neural basis of speech and language networks in healthy controls and clinical populations. Since no review of brain networks is complete without an understanding of the assumptions underlying connectivity analyses and the relationship between regions implicated within a network, we present an overview of the methods used to study interacting brain regions observed in neuroimaging under passive and active task conditions. Finally, we illustrate the application of these methods to data from one of our MEG studies, showing how effective connectivity analyses can help to interpret results in neuroimaging.

8.2 Applications of Causal Models in Studies of Speech and Language

With the advent of non-invasive functional neuroimaging methods in the late 1970s, localization theories of language—based on brain lesion studies—have given way to distributed models of language, clearly implicating a network of sequential and parallel functional connections. Speech and language processing rely on distinct structural and functional pathways woven together in a complex network. This renders processing within the network well suited to effective connectivity analysis using causal modeling. Despite the large number of studies examining various components of the language system (e.g., orthography, phonology, morphology, semantics, syntax), the relationship between these processes and the direction of the interactions between them remain far less understood. This is particularly relevant in early childhood, when mastery of language entails competition between various developing linguistic processes within the speech network. The relative efficiency of these processes (Wehner et al. 2007; Mody et al. 2008; Han et al. 2012) may well dictate the pattern of effective connectivity during information flow between regions in the network. Additionally, the application of causal modeling to disorders of speech and language may provide valuable insights not just about which connections are affected but also about their role and plasticity in the recovery process following intervention. In this section, we review a few studies that apply DCM and GC to questions about the functional organization of speech and language processes in the brain. The studies cited are not intended to present an exhaustive review of this topic but rather to stimulate ideas for new research using these methods.

8.2.1 Speech Perception

The field of speech perception encompasses a wide range of theoretical approaches to fundamental issues which affect our understanding of the representation and processing of speech. These include questions about mechanisms underlying our capacity to recover linguistic units like phonemes, syllables, and words from the speech signal, the neurobiological basis of speech perception, and the interface between perception and lexical influences. While the mechanism by which we analyze the incoming acoustic–phonetic signal to recover linguistic form and meaning remains controversial (Studdert-Kennedy & Mody, 1995), creative experimental paradigms in combination with neuroimaging methods and analysis approaches have allowed us to probe the relationships between sensory and linguistic processes aimed at better understanding language processing. For example, it has been shown that when presented with an ambiguous phoneme (e.g., s/ʃ) in a lexical context (e.g., /shampoo/), subjects tend to perceive this sound as /ʃ/, consistent with its context. Interpretation of the latter as a word, the phenomenon known as the Ganong effect (Ganong 1980). Two competing hypotheses have been proposed to explain the Ganong effect. Some researchers argue that lexical knowledge may influence speech categorization through a top-down mechanism, while others believe the Ganong effect may be mediated through a feedforward mechanism and lexical knowledge is integrated at a post-perceptual stage. Gow et al. (2008) tested these two hypotheses using GC with MEG and EEG data collected from healthy subjects. During the experiment, subjects listened to words and were asked to make a decision whether the words start with the sound [s] or [ʃ]. Causal interactions were studied between nine regions of interest (ROIs): left posterior superior temporal gyrus (pSTG), right superior temporal gyrus, right medial wall of temporal lobe, left supramarginal gyrus (SMG), left angular gyrus, left anterior superior temporal gyrus, left inferior frontal gyrus (IFG), left anterior cingulate gyrus, and left posterior cingulate gyrus. The GC analysis suggested that information flowed from left SMG to left pSTG during the 280–480 ms time period when greater left pSTG activation associated with the Ganong effect was observed. In brief, these findings support top-down influences of lexical knowledge on phonetic processing (Gow et al. 2008).

Priming paradigms have also been used to examine phonological-semantic relations in speech perception. Priming a word with another semantically or phonologically related one usually results in faster reaction time and reduced brain response for the target. In a MEG study, Kujala et al. (2012) tested the hypothesis that accurate and faster responses may be associated with more efficient interactions between brain regions. Subjects read lists of four words and were instructed to press a button when they found that the same word was presented twice in a list. The last word in the list was primed semantically or phonologically by the first three words. Enhanced coherence between relevant language areas was associated with the priming effect. In addition, GC analysis indicated significant connectivity from the left superior temporal to the left occipito-temporal cortex.

Interestingly, previous studies have also found silent reading to activate the auditory cortex. However, it is unclear whether the auditory regions were involved through bottom-up inputs from the visual regions or top-down mechanisms. Perrone-Bertolotti et al. (2012) addressed this question using EEG and GC. Intracerebral EEG signals were recorded from four patients with epilepsy with implanted EEG electrodes. During the experiment, the patients were presented two stories one word at a time and instructed to read the words written in gray and ignore the ones in white. The authors observed activation of the voice-selective regions during silent reading in the attention but not in the ignore condition. GC was used to study the interactions between the visual and auditory cortices. The results showed greater connectivity from the visual cortex (ventral occipital temporal cortex) to the primary auditory cortex than from the primary auditory cortex to the visual cortex when subjects attended to the words (Perrone-Bertolotti et al. 2012).

As with GC, studies using DCM have also yielded interesting insights about causal interactions during language processing. Heim et al. (2009b) identified three brain regions including left Brodmann area (BA) 44, inferior temporal gyrus (ITG), and BA 45 in fMRI during a phonological decision task (PDT; "Does the stimulus begin with a fricative or stop consonant?") and a lexical decision task (LDT; "Is the stimulus a word or a pseudoword?"). Four alternative models of varying interaction patterns between these areas—in relation to each of the tasks—were tested and compared using DCM. Using Bayesian model comparison, the best model found LDT and PDT modulate the connection from ITG to BA 45. Statistical analyses of the parameter estimates of this model also suggested a positive modulation of the connection from ITG to BA 45 by LDT in keeping with BA 45's role in explicit lexical decision making (Heim et al. 2009b). In a follow-up study of BA 44 and BA 45 using a verbal fluency task, DCM revealed that although both regions contributed to the phonological and semantic verbal fluency tasks, there was a functional distinction between the regions: BA 45 was associated with lexical retrieval while BA 44 supported the processing of phonological information in these tasks (Heim et al. 2009a).

8.2.2 Speech Production

Producing speech is a complex act of planning and coordination of articulatory gestures toward linguistic goals. Eickhoff et al. (2009) identified a core brain network consistently involved in overt speech production based on their own work and a meta-analysis of 19 other studies. The network consisted of BA 44, the anterior insula, cerebellum, basal ganglia, ventral premotor cortex, primary motor cortex. They further examined the causal interactions within this network by testing four different hypothetical models. In all the models, speech-specific inputs entered via BA 44. Model 1 postulated that BA 44 projected to the insula, which relayed the signals to the cerebellum and basal ganglia. Signals would then be sent to the premotor cortex and finally arrive at the primary motor cortex. Model 2 assumed that information flows in parallel from BA 44 to the cerebellum, basal ganglia, and insula. Signals from these three regions were then integrated in the premotor cortex and sent to primary motor cortex. As for model 3 and model 4, signals from BA 44 were first sent to the insula which then relayed these signals in parallel to the cerebellum, basal ganglia, and premotor cortex. However, model 3 postulated cerebellum and basal ganglia projected directly to primary motor cortex; whereas model 4 postulated signals from cerebellum and basal ganglia were integrated at the premotor cortex before being sent to the primary motor cortex. Results from Bayesian model selection procedure suggested model 1 to be the optimal model representing the functional architecture of overt speech production. The speech production network appeared to have two subsystems. The connections from insula to cerebellum and basal ganglia were associated with condition-specific inputs, whereas the effective connectivity from cerebellum and basal ganglia to premotor cortex was correlated with the number of the words the subjects produced (Eickhoff et al. 2009). More recently, Holland et al. investigated how anodal transcranial direct current stimulation (tDCS) modulated causal interactions within the speech network (Holland et al. 2016). Ten healthy controls underwent fMRI during an overt picture naming task while anodal tDCS or sham stimulation was delivered to the left frontal area. Compared to sham stimulation, anodal tDCS was shown to be associated with faster naming responses. The causal interactions between two frontal nodes, inferior frontal sulcus (IFS), and ventral premotor cortex (VPM) were studied. The DCM analysis revealed excitatory influences from VPM to IFS and inhibitory influences from IFS to VPM. In addition, anodal tDCS elicited stronger inhibitory influences from IFS to VPM while sham stimulation elicited stronger self-inhibitory influences in VPM.

Whereas these and other studies clearly implicate inferior frontal areas in speech production, the exact role of Broca's area during word production remains unclear.

Flinker et al. (2015) elucidated the neural substrates of overt word production using intracranial electrocorticographic recordings in patients with refractory epilepsy. Subjects participated in a word repetition task and a word-reading task. The authors studied the causal interactions between the auditory cortices (STG and superior temporal sulcus, STS), Broca's area, and the motor area using Event-Related Causality (ERC), an extension of GC for multichannel data. They found propagation of signals from auditory cortices to Broca's area and then to the motor system during speech production. They also observed that Broca's area seemed to be activated only during pre-articulatory stages and suspended during spoken responses. ERC analysis revealed that signals were sent from STG to Broca's area in the first 200 ms and sent back later from Broca's area to STG. The authors thought this interaction pattern may be associated with phonological representation of words. Broca's area also projected to the motor areas but the influence was not seen during articulation (Flinker et al. 2015).

8.2.3 Speech Comprehension

Speech comprehension provides a useful framework within which to examine processes of speech perception and production. At a basic level, it raises questions about the interactions between sensory and motor areas. Londei and colleagues combined independent component analysis (ICA) and GC to study causal interactions between sensory and motor areas during speech comprehension (Londei et al. 2010). In an fMRI experiment, subjects listened to words, pseudo-words, and reversed-words (words and pseudo-words played in reverse). The GC results showed that the tasks had prominent modulatory effects on the causal interactions between ICA-derived difference maps. Specifically, the directed functional connections from the somatosensory map to the motor map and from the inferior frontal gyrus/inferior parietal lobule area to the somatosensory area were only seen in the words and pseudo-words conditions. Similar modulatory effects were also observed for the causal interactions between the cerebellum, the motor cortex, and the inferior frontal gyrus/inferior parietal lobule. In a separate study, Yang and Shu (2012) addressed the question whether the activation of the premotor and primary motor cortices seen in action verb comprehension is caused by motor processes supporting semantic access of action verbs or by action language comprehension. Subjects underwent fMRI scanning, while they passively viewed action verbs during the experiment. GC analysis was performed to study the interactions between 3 ROIs including left posterior middle temporal gyrus (MTG), primary motor (M1), and premotor cortex. The MTG (representing the region for lexical semantic processing) was selected based on brain activation during the passive verb reading task, whereas the M1 and premotor cortex were identified based on activation during passive reading within a hand motion task involving pantomiming a grasping action. The results showed bidirectional functional connectivity between MTG and premotor cortex and unidirectional connection from MTG to M1. These results suggest that the premotor and primary motor cortices might function differently: the premotor cortex may interact with the MTG and be associated with motor simulation in action language comprehension; the motor cortex may contribute to motor processes mediating semantic access of action verbs (Yang and Shu 2012).

Researchers have also attempted to integrate findings from functional neuroimaging of speech comprehension with studies of brain structural connectivity to further understand the functional anatomy of speech and language. In one such study, Saur and colleagues (2010) collected fMRI data from 33 subjects doing an auditory sentence comprehension task. Subjects listened to meaningful speech (SP), pseudo speech (PS), and reversed speech (REV) and were instructed to press a button at the end of the stimulus regardless of which kind of speech they heard. A directed partial correlation (dPC) method was used to study GC. Causal interactions within four brain networks identified by contrasting PS with REV and SP with PS were investigated. For any significant causal connection, probabilistic tractography was performed with diffusion tensor imaging data to track fiber pathways between the corresponding nodes. For the phonological network, a dorsal route (superior longitudinal fasciculus and arcuate fasciculus) connecting the temporal and premotor regions and a ventral route (extreme capsule, EmC) connecting the temporal and prefrontal regions were found. In the left and right semantic networks, interactions between the temporal and prefrontal regions were also mediated through EmC. In the bilateral semantic network, commissural fibers seemed to mediate the interhemispheric interactions (Saur et al. 2010).

Noppeney and colleagues (2006) examined language comprehension in healthy subjects during a one-back task using fMRI. Pictures and words (spoken or written) of animals and tools were presented, and subjects were asked to indicate whether the stimuli were identical, represented a similar action, or had similar size in real life. The experiment was designed to provide evidence on how category-selective responses were modulated by modality (verbal vs. non-verbal) and task (implicit vs. explicit semantic task). The authors used DCM to investigate the causal interactions between five cortical regions involved in this paradigm: superior temporal, occipital, prefrontal, anterior inferior parietal, and fusiform regions. The results suggested that modality-dependent category-selective responses were mediated through a bottom-up (from occipital cortex to fusiform and anterior inferior parietal cortex) mechanism, whereas task-dependent category-selective responses were mediated through a top-down (from prefrontal to fusiform and anterior inferior parietal regions) strategy (Noppeney et al. 2006).

Taken together, GC and DCM findings from studies of speech perception, production, and comprehension provide further evidence of the dynamic and interactive nature of speech and language processing. They highlight the complexity of higher level cognitive processing by revealing the varying patterns of directed connectivity between and within implicated brain regions. Creative experimental paradigms that draw on the interactions between language processes are helping advance our understanding of the causal influences underlying the execution of speech comprehension and production.

8.2.4 *Clinical Studies*

Causal models have also been used in clinical studies of language processing in patients with brain damage following a stroke. In one such study, Seghier and colleagues compared a patient with left putamen damage to 24 healthy control subjects during reading aloud words and a picture naming task using fMRI (Seghier et al. 2014). The patient performed well on both tasks despite the important role the putamen plays in speech production, raising questions about a potential compensatory mechanism. Findings from the study suggested a compensatory area located in the left premotor cortex that exhibited great activation in the patient when reading aloud and naming the pictures. DCM was then used to identify neural pathways through which the compensatory area may function to support speech production in the patient. Given that connectivity of the putamen was impaired (because this region was damaged by the stroke), reading aloud words and naming pictures appeared to induce greater changes in connections from several other regions to the premotor area including the thalamus, ventral occipital cortex, anterior occipito-temporal cortex, and motor cortex (Seghier et al. 2014). In a separate study, Kiran and her team (2015) also used DCM to investigate the neural basis of language recovery, though specifically with regard to rehabilitation effects in patients with aphasia. Eight patients underwent fMRI scanning during a picture naming task and semantic feature matching task before and after a semantic feature-based naming therapy. Activation of several regions associated with language processing was enhanced following treatment and coincided with improvement in behavioral performance. DCM analysis showed connectivity of the LIFG was significantly modulated after rehabilitation consistent with its established role in speech and language (Kiran et al. 2015). These findings further reinforce the importance of causal modeling approaches in understanding the neural basis of speech and language processing in health and disease. While we have reviewed only a small sample of speech research that exploits causal models, the studies make it clear that such models have a lot to offer to cognitive neuroscience. Below we describe the some of the more commonly used causal modeling methods.

8.3 Models of Causal Brain Connectivity

Researchers have developed several models to study causal interactions among distributed brain regions involved in cognitive processing. The models have helped elucidate the functional significance of various networks and their anatomical underpinnings. We focus here on GC and DCM, the two most widely used causal models in (directed) brain connectivity studies.

8.3.1 Granger Causality

GC, first introduced in the field of econometrics (Granger 1969), was applied to electrophysiological recordings in 1999 (Bernasconi and Konig 1999). In 2003, Goebel and colleagues applied GC to fMRI data to study directed interactions within a brain network activated during a visuo-motor mapping task (Goebel et al. 2003). GC directly models the causal interactions of two brain regions (region 1 and region 2) using the observed (BOLD, EEG, or MEG) signals $y_1[n]$ and $y_2[n]$ of these regions. GC is usually implemented using an autoregressive model:

$$y_1[n] = -\sum_{i=1}^{p} A(i) y_1[n-i] + \sum_{i=1}^{p} B(i) y_2[n-i] + \varepsilon[n], \tag{8.1}$$

where $y_1[n]$ represents the observed signal of region 1 at the current time point (n) while $y_1[n-i]$ and $y_2[n-i]$ denote the observed signals of region 1 and region 2 in the past p time points. $A(i)$ and $B(i)$ are the autoregressive model parameters, and $\varepsilon[n]$ is the prediction error. Region 2 is considered to have causal effects on region 1 when we can better predict $y_1[n]$ using the past values of both regions ($y_1[n-i]$ and $y_2[n-i]$) than using only the past signal of region 1 ($y_1[n-i]$).

It is important to note that the use of GC to model fMRI data has some disadvantages. Strictly speaking, GC is a model of directed functional connectivity because it is not based upon a model of connectivity per se. Note that the autoregressive model above does not contain a unique parameter that describes the coupling between the two regions. In other words, it is a model of statistical dependencies not connectivity per se. Moreover, these statistical dependencies are among measured signals—as opposed to the neuronal activity causing measured signals. In electrophysiology, this is generally not a serious problem; however, the situation is more complicated in fMRI: it is well known that blood oxygenation level dependent (BOLD) signals do not directly measure neuronal activities. Instead, BOLD signals are hemodynamic convolutions of underlying neuronal activity (Fig. 8.1). Thus, the causal influences that the activity of one region exerts on another, inferred directly from BOLD signals, may not reflect the coupling mediated at the neuronal level, especially when the hemodynamic process differs significantly over different regions. For example, there might be a differential lag and dispersion in the fMRI signal—due solely to the sluggishness of the hemodynamic response—that reverses temporal precedence at the neuronal level. In the most extreme case, this can, in principle, completely reverse the direction of the inferred functional connectivity between regions. It has been suggested that "the inevitable effects that the hemodynamic system imposes on causal inferences in fMRI data, lead us toward the methods in which causal inferences can take place in latent neuronal level, rather than observed BOLD time series" (Kadkhodaeian Bakhtiari and Hossein-Zadeh 2012). One way to address this shortcoming of GC is to obtain predictions of BOLD responses by exploiting distinct models of neuronal activity and hemodynamics. Then, by jointly optimizing these models using fMRI data, one obtains region-spe-

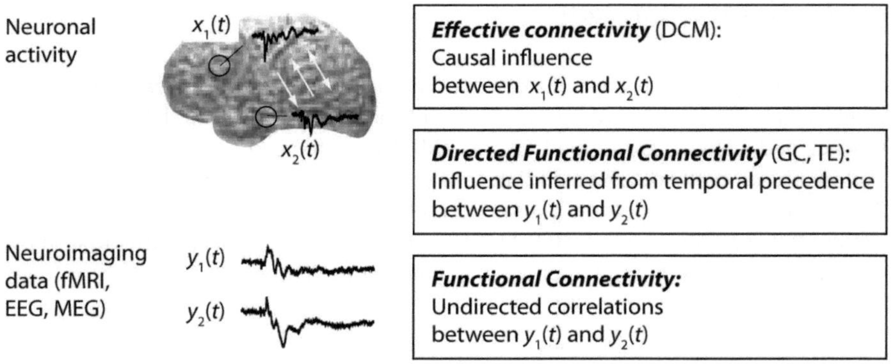

Fig. 8.1 Schematic illustration of functional, directed, and effective connectivity

cific estimates of hemodynamic parameters that can accommodate regional variations in the hemodynamic response functions, while at the same time providing estimates of effective connectivity. This is the approach taken by DCM as described below.

8.3.2 Dynamic Causal Modeling

Basic Principles of DCM

The basic idea behind DCM is to treat the brain as a nonlinear input–output system (Friston et al. 2003; Stephan et al. 2010; Pinotsis and Friston 2014). DCM tries to infer causal interactions between different brain regions at a neuronal level using neuronal state equations:

$$\dot{z} = F\left(z,u,\theta\right), \tag{8.2}$$

where neuronal states $\dot{z}$ represents the neuronal activities of all the regions within a specific network. Changes in neuronal activity of each region in this network are assumed to be a nonlinear function F comprising three sets of variables: the neuronal states of any other interacting regions (z), experimental inputs (u), and the strength of the effective connectivity (θ). In other words, DCM characterizes causal interactions between brain regions in terms of parameters θ that determine the form and strength of influences among these regions. These parameters usually include time constants or synaptic strengths of the connections between brain regions. The mathematical form of these dependencies $F = F(z, u, \theta)$ and the pattern of allowable connections represent the structure of the brain network. Each node in the network or region is perturbed by some endogenous or subcortical input u. By integrating Eq. (8.2) in time, one obtains predictions of the time series of neural activity in each region. Fitting these predictions to empirical data allows one to obtain estimates of effective connectivity between these regions.

Usually there exist several competing hypotheses about the mechanisms or architecture underlying a cognitive process—and one may want to compare these hypotheses given the observed data. This can be accomplished using a Bayesian model selection procedure. That is, given the data y, the model evidence of a specific model m can be described as

$$p(y|m) = \int p(y|\theta,m)\, p(\theta|m)\, d\theta. \tag{8.3}$$

The model evidence is the probability of obtaining the data given a model. It can be approximated based on a free energy function that is obtained by fitting model predictions to observed responses using a variational Bayesian algorithm (i.e., approximate Bayesian inference). Given approximate model evidence for competing models, the Bayes factor B_{ij} can be used to decide if model m_i is better than m_j:

$$B_{ij} = \frac{p(y|m_i)}{p(y|m_j)}. \tag{8.4}$$

If $B_{ij}>20$, we can say that model m_i is better than m_j. Or, more rigorously, the evidence for the better model is 20 times that of the alternative model. Generally, one works with log evidences, which means that one can compare multiple models at the same time in terms of their relative log evidence.

DCM for fMRI

DCM for fMRI was first developed in 2003 (Friston et al. 2003) and employs a bilinear approximation to any neuronal dynamics described by Eq. (8.2):

$$\dot{z} \approx Az + \sum u_j B^j z + Cu, \tag{8.5}$$

where the matrix A incorporates anatomical information and represents the average connectivity from one region to another in the absence of external inputs. The external inputs (e.g., visual or auditory stimuli used in an experiment) are usually assumed to enter sensory regions and cause condition-specific responses in these regions (corresponding to the matrix C in Eq. 8.5). In addition, inputs u may also induce changes in average connectivity A—these are modulatory effects of experimental inputs and are described by the (bilinear) matrix B in Eq. (8.5). Recently, we have extended this traditional deterministic DCM to stochastic DCM which accommodates random fluctuations in hidden neuronal and physiological states (Li et al. 2011). With stochastic DCM, one is also allowed to model the effective connectivity among brain regions using resting-state fMRI data (Li et al. 2012).

DCM for EEG/MEG

DCM for EEG/MEG is based on nonlinear models of neuronal activity that often include details about the biophysical properties of cortical sources. These models exploit the high temporal resolution of EEG/MEG data and come in several flavors including DCM for evoked responses (ERPs), DCM for induced responses, DCM for cross spectral density, DCM with neural fields, DCM for phase coupling, etc. (Kiebel et al. 2006; Pinotsis et al. 2012; Moran et al. 2013; Pinotsis et al. 2013). DCM for ERPs and DCM for induced responses are among the most widely used among these.

DCM for ERPs is based on detailed physiological models of causal interactions. It uses realistic physiological neural models that exploit EEG/MEG data to explain interactions among brain sources at the neuronal level (Kiebel et al. 2006; Kiebel et al. 2007). Local neural populations can be thought of as forming microcircuits or (macro) columns of the size of a few millimeters. In this context, each cortical source corresponds to a local microcircuit comprising neuronal populations that are usually assigned to different layers. These are connected via extrinsic (between-source) connections that can be either forward or backward. Superficial pyramidal cells are thought to be the sources of forward connections that—under predictive coding models of cortical hierarchies—convey prediction errors to regions higher in the cortical hierarchy, while backward connections are thought to mediate predictions that are conveyed to lower areas from deep pyramidal populations. Local populations within a cortical source are connected according to (intrinsic) connectivity rules that follow empirical studies of cortical anatomy and physiology (Douglas and Martin 1991). The proposed microcircuitry, as used in DCM, is known as the canonical microcircuit (Bastos et al. 2012; Pinotsis et al. 2014). This neural model comprises superficial and deep pyramidal cells in layers in 2/3 and 5, respectively, spiny stellate cells that receive input in layer 3 and inhibitory interneurons that are interspersed across cortical layers.

In DCM, the depolarization of the pyramidal cell population is the presumed source of EEG/MEG signals (David et al. 2006). However, biophysical computational modeling of the dendritic currents in the pyramidal cells, resulting from different spatial distributions of synaptic inputs, could provide a more detailed model for the generation of the EEG and MEG signals (Ahlfors and Wreh 2015; Ahlfors et al. 2015).

DCM for induced responses, in contrast, employs a phenomenological as opposed to a physiological model of brain connectivity (Chen et al. 2008). DCM for induced responses models the influence one region exerts on another in the frequency domain (as opposed to time domain in the case of DCM for ERPs above), based on time-frequency analysis of EEG/MEG data. Changes in the activity of a region are assumed to be caused by activity (summarized as time evolving power) of brain regions that are connected to it, and experimental inputs entering this region.

8.3.3 *Other Measures of Causal Connectivity*

As mentioned earlier, DCM and GC are among the most widely used models in studies of causal connectivity. Box 1 presents a brief overview of other approaches for directed connectivity analysis. These include psychophysiological interactions (PPI), structural equation modeling (SEM), transfer entropy (TE), and phase slope index (PSI).

Box 1 Models to Study Causal Interactions among Distributed Brain Regions

- Psychophysiological interactions (PPI) (Friston et al. 1997): The basic idea behind PPI is that the slope of the regression of the activity of region 1 on that of region 2 is considered the contribution of region 2 to region 1, and as such can be related to effective connectivity between these regions (under a very simple linear model of neuronal coupling).
- Structural equation modeling (SEM) (Mclntosh and Gonzalez-Lima 1994): SEM differs from PPI in that it can be used to study the causal interactions among multiple brain regions simultaneously. It uses an anatomical model that embodies directed connections between distant regions, usually based on anatomical knowledge.
- Transfer entropy (TE): This method is based on transition probabilities and describes information exchange—in a statistical sense—among regions (Schreiber 2000). TE is capable of detecting nonlinear causal interactions and deals with signal cross-talk, which is crucial for the analysis of EEG and MEG data (Vicente et al. 2011).
- Phase slope index (PSI): PSI is defined as a weighted average of the phase slope that accommodates different frequencies while insensitive to independent sources and confounding factors. The basic idea behind PSI is that a positive phase slope would be seen in the cross-spectra between two time courses y_1 and y_2, if y_1 drives y_2 (Nolte et al. 2008).

8.4 Example of DCM Application to MEG

In this section, we present some preliminary results illustrating the use of DCM with MEG data from a task used in one of our previous studies (Wehner et al. 2007). The findings reveal the exciting potential of this approach to probe the brain's functional architecture and obtain a deeper understanding of the neurobiology of speech and language.

8.4.1 Task

We collected MEG data from a healthy adult subject performing a homophone judgment task. During the experiment the subject saw pairs of words: the first word presented for 300 ms, followed by a second word that was either semantically related (the synonym foil condition, e.g., jet and plane), a homophone (the homophone condition, e.g., plain and plane), or unrelated (the control condition, e.g., dog and plane). The subject's task was to respond with the appropriate button press "yes" to homophone word pairs, "no" otherwise). For more details on the stimuli and experimental procedure please refer to our previous work (Wehner et al. 2007). The subject exhibited a semantic interference effect (SIE), i.e., a longer reaction time to reject stimulus pairs containing a synonym foils compared to unrelated control stimulus pairs, similar to the findings in our previous study. This effect suggests that both semantic and phonological systems appear to be activated during single word reading. In this section, we will use DCM to ask whether the SIE effect is mediated through top-down or bottom-up mechanisms—as reflected in the interactions between phonological and semantic regions of interest.

8.4.2 MEG Data Collection and Preprocessing

MEG data was collected at the Athinoula A. Martinos Center for Biomedical Imaging using a 306-channel MEG system (VectorView, ElektaNeuromag). The MNE software (http://martinos.org/mne/stable/index.html) was used for data preprocessing and to obtain distributed source estimates of the cortical activity (for complete details, please refer to Wehner et al. 2007). Five regions involved in word processing including the visual cortex (the occipital lobe, OCC), fusiform gyrus (FFG), STS, IFG, and prefrontal cortex (PFC)/insula were identified (Fig. 8.2).

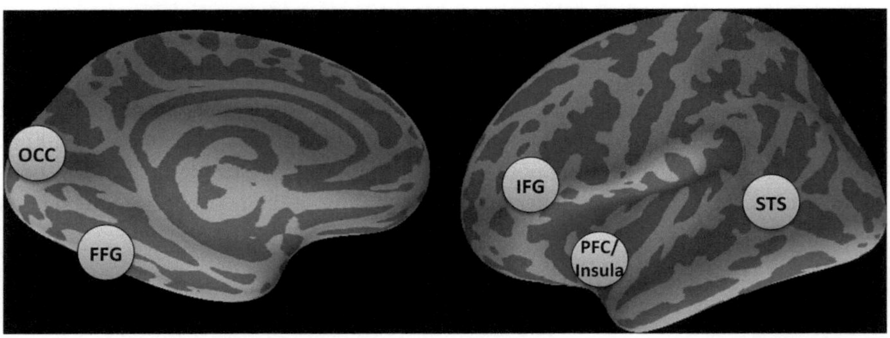

Fig. 8.2 The locations of the five regions of interest

Averaged source time courses for the selected regions of interest were extracted and then converted to SPM format (http://www.fil.ion.ucl.ac.uk/spm/software/spm8/).

8.4.3 Model Specification and Inversion

In this study, we are interested in modeling the differences in effective connectivity between the control and synonym conditions. We assume that the experimental inputs first enter the OCC. Signals are then sent to the FFG from where they are relayed to IFG and STS. IFG and STS are hypothesized to interact reciprocally with the signal finally sent to the PFC/insula, the decision endpoint. We tested three competing hypotheses that may account for the differences in brain activity between the control and synonym conditions: (1) reciprocal connections between IFG and STS; (2) connections from STS to IFG and from IFG to PFC/insula; or (3) connections from IFG to STS and from STS to PFC/insula. These competing hypotheses about the functional architecture of the language network correspond to the three models shown in Fig. 8.3. We used a local field potentials (LFP) neural mass model within DCM to explain the estimated source time courses for each region in the hypothetical models above.

8.4.4 Bayesian Model Selection

After model inversion, Bayesian model selection procedures were used to evaluate the three models according to the model evidence using a fixed-effects (FFX) approach. The results point to model 3 as the winning model with the highest model evidence (Fig. 8.4), suggesting that the differences in brain activity between the control and synonym conditions may have been mediated through the connections from the IFG to STS and from the STS to PFC/insula.

Although we have skipped over many of the subtleties of DCM and model comparison, this simple result illustrates the utility of dynamic causal modeling when there are well-posed questions about the functional anatomy of speech and language.

In essence, we have been able to disambiguate among equally plausible architectures—each entailing multiple processing routes—using standard electromagnetic responses and Bayesian model comparison. Note that the underlying DCM is just like a standard forward model of electromagnetic responses (based upon equivalent current dipoles). However, here, our knowledge of the language network has enabled us to equip this model with directed connections, while standard Bayesian modeling techniques enable us to assess the evidence for competing hypotheses, cast as different patterns of directed connections.

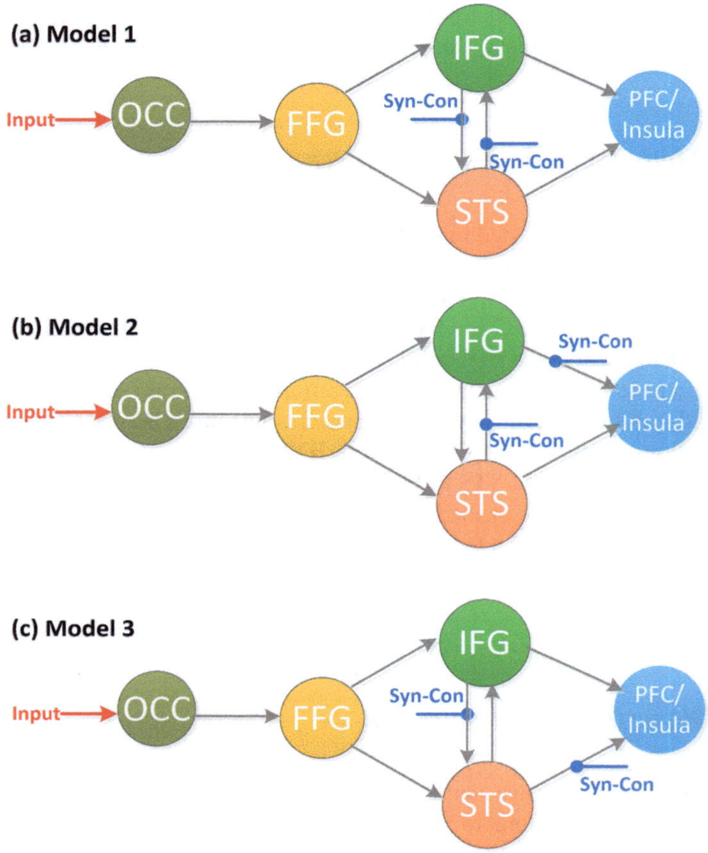

Fig. 8.3 Three hypotheses representative of the potential causal interactions between regions within the language network during a semantic interference task "Syn-Con" indicates hypothesized connections that may underlie the observed differences between Synonym and Control conditions

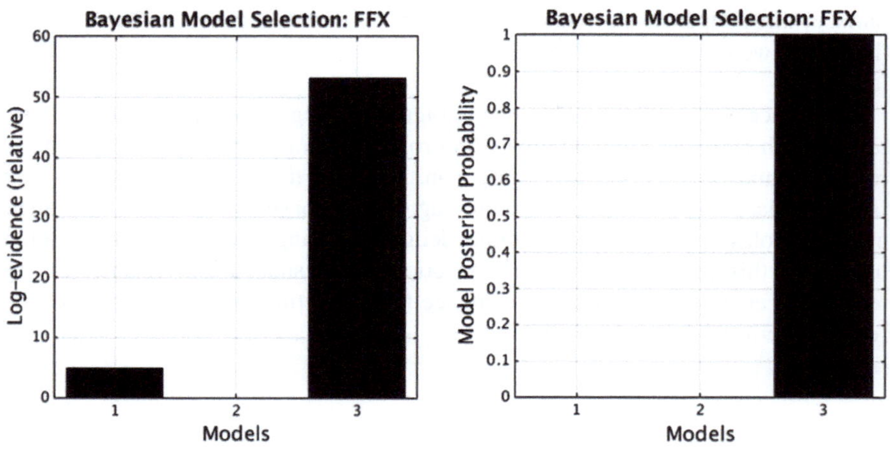

Fig. 8.4 The results of Bayesian model selection

8.5 Strengths and Weaknesses of DCM and GC

Although GC and DCM methods are based on different basic principles they are not competitive but complementary as they attempt to answer distinct research questions (Friston et al. 2013; Seth et al. 2015). Here, we summarize some important differences between the two methods: (1) GC operates directly at the observed BOLD or EEG/MEG signal level, thus GC measures directed functional connectivity. In contrast, DCM models the causal interactions between different regions at the hidden neuronal level, and measures the influence neuronal activity in one region exerts over another, i.e., effective connectivity; (2) GC and DCM treat experimental inputs in different ways. GC does not take into account the experimental inputs when investigating the interactions between different regions, whereas DCM explicitly models the inputs in the neuronal state equations. Experimental inputs are assumed to enter into sensory nodes and directly change neural activity, effectively leading to dynamic perturbations of the network response. Experimental inputs can also modulate specific brain connections—amounting to targeted perturbations of the cortical network. This allows one to investigate how different experimental conditions modulate effective connectivity; (3) Due to the lack of a hemodynamic model, one should be careful when utilizing GC in fMRI, especially when there are significant differences in regional hemodynamic latency. (4) DCM allows one to compare alternative (mechanistic) hypotheses about the neural correlates of observed brain responses; for example, by comparing alternative networks using Bayesian model selection, one can find the functional architecture of a network of regions implicated in overt speech production.

8.6 Conclusion

Over the last decade, there has been increasing evidence for a distinct speech network comprising distant but interacting brain regions, including the STG, IFG, SMG, premotor cortex, primary motor cortex, insula, and cerebellum. However, the direction of information flow within the network has remained a challenge, eluding questions about top-down vs. bottom-up relations between interacting processes. Recent advances in analysis methods using Dynamic Causal Modeling are beginning to address these questions. Clinical applications of these approaches focused on mechanisms mediating speech and language recovery and rehabilitation are helping provide important insights about the structural and functional architecture of the human brain in health and disease.

References

Ahlfors, S. P., Jones, S. R., Ahveninen, J., Hamalainen, M. S., Belliveau, J. W., & Bar, M. (2015). Direction of magnetoencephalography sources associated with feedback and feedforward contributions in a visual object recognition task. *Neuroscience Letters, 585*, 149–154. doi:10.1016/j.neulet.2014.11.029

Ahlfors, S. P., & Wreh, C., II. (2015). Modeling the effect of dendritic input location on MEG and EEG source dipoles. *Medical & Biological Engineering & Computing, 53*(9), 879–887. doi:10.1007/s11517-015-1296-5

Bastos, A. M., Usrey, W. M., Adams, R. A., Mangun, G. R., Fries, P., & Friston, K. J. (2012). Canonical microcircuits for predictive coding. *Neuron, 76*, 695–711. doi:10.1016/j.neuron.2012.10.038

Bernasconi, C., & Konig, P. (1999). On the directionality of cortical interactions studied by structural analysis of electrophysiological recordings. *Biological Cybernetics, 81*, 199–210. doi:10.1007/s004220050556

Chen, C. C., Kiebel, S. J., & Friston, K. J. (2008). Dynamic causal modelling of induced responses. *NeuroImage, 41*, 1293–1312. doi:10.1016/j.neuroimage.2008.03.026

David, O., Kiebel, S. J., Harrison, L. M., Mattout, J., Kilner, J. M., & Friston, K. J. (2006). Dynamic causal modeling of evoked responses in EEG and MEG. *NeuroImage, 30*, 1255–1272. doi:10.1016/j.neuroimage.2005.10.045

Douglas, R. J., & Martin, K. A. (1991). A functional microcircuit for cat visual cortex. *The Journal of Physiology, 440*, 735–769.

Eickhoff, S. B., Heim, S., Zilles, K., & Amunts, K. (2009). A systems perspective on the effective connectivity of overt speech production. *Philosophical Transactions. Series A, Mathematical, Physical, and Engineering Sciences, 367*, 2399–2421. doi:10.1098/rsta.2008.0287

Flinker, A., Korzeniewska, A., Shestyuk, A. Y., Franaszczuk, P. J., Dronkers, N. F., Knight, R. T., et al. (2015). Redefining the role of Broca's area in speech. *Proceedings of the National Academy of Sciences of the United States of America, 112*, 2871–2875. doi:10.1073/pnas.1414491112

Friston, K., Moran, R., & Seth, A. K. (2013). Analysing connectivity with Granger causality and dynamic causal modelling. *Current Opinion in Neurobiology, 23*, 172–178. doi:10.1016/j.conb.2012.11.010

Friston, K. J. (1994). Functional and effective connectivity in neuroimaging: A synthesis. *Human Brain Mapping, 2*, 56–78.

Friston, K. J., Buechel, C., Fink, G. R., Morris, J., Rolls, E., & Dolan, R. J. (1997). Psychophysiological and modulatory interactions in neuroimaging. *NeuroImage, 6*, 218–229. doi:10.1006/nimg.1997.0291

Friston, K. J., Harrison, L., & Penny, W. (2003). Dynamic causal modelling. *NeuroImage, 19*, 1273–1302.

Ganong, W. F., III. (1980). Phonetic categorization in auditory word perception. *Journal of Experimental Psychology. Human Perception and Performance, 6*, 110–125.

Goebel, R., Roebroeck, A., Kim, D. S., & Formisano, E. (2003). Investigating directed cortical interactions in time-resolved fMRI data using vector autoregressive modeling and Granger causality mapping. *Magnetic Resonance Imaging, 21*, 1251–1261.

Gow, D. W., Jr., Segawa, J. A., Ahlfors, S. P., & Lin, F. H. (2008). Lexical influences on speech perception: A Granger causality analysis of MEG and EEG source estimates. *NeuroImage, 43*, 614–623. doi:10.1016/j.neuroimage.2008.07.027

Granger, C. (1969). Investigating causal relations by econometric models and cross-spectral methods. *Econometrica, 37*, 424–438.

Han, J., Mody, M., & Ahlfors, S. P. (2012). Gamma phase locking modulated by phonological contrast during auditory comprehension in reading disability. *Neuroreport, 23*, 851–856. doi:10.1097/WNR.0b013e32835818e1

Heim, S., Eickhoff, S. B., & Amunts, K. (2009a). Different roles of cytoarchitectonic BA 44 and BA 45 in phonological and semantic verbal fluency as revealed by dynamic causal modelling. *NeuroImage, 48*, 616–624. doi:10.1016/j.neuroimage.2009.06.044

Heim, S., Eickhoff, S. B., Ischebeck, A. K., Friederici, A. D., Stephan, K. E., & Amunts, K. (2009b). Effective connectivity of the left BA 44, BA 45, and inferior temporal gyrus during lexical and phonological decisions identified with DCM. *Human Brain Mapping, 30*, 392–402. doi:10.1002/hbm.20512

Holland, R., Leff, A. P., Penny, W. D., Rothwell, J. C., & Crinion, J. (2016). Modulation of frontal effective connectivity during speech. *NeuroImage, 40*, 126–133. doi:10.1016/j.neuroimage.2016.01.037

Kadkhodaeian Bakhtiari, S., & Hossein-Zadeh, G. A. (2012). Subspace-based identification algorithm for characterizing causal networks in resting brain. *NeuroImage, 60*, 1236–1249. doi:10.1016/j.neuroimage.2011.12.075

Kiebel, S. J., David, O., & Friston, K. J. (2006). Dynamic causal modelling of evoked responses in EEG/MEG with lead field parameterization. *NeuroImage, 30*, 1273–1284. doi:10.1016/j.neuroimage.2005.12.055

Kiebel, S. J., Garrido, M. I., & Friston, K. J. (2007). Dynamic causal modelling of evoked responses: The role of intrinsic connections. *NeuroImage, 36*, 332–345. doi:10.1016/j.neuroimage.2007.02.046

Kiran, S., Meier, E. L., Kapse, K. J., & Glynn, P. A. (2015). Changes in task-based effective connectivity in language networks following rehabilitation in post-stroke patients with aphasia. *Frontiers in Human Neuroscience, 9*, 316. doi:10.3389/fnhum.2015.00316

Kujala, J., Vartiainen, J., Laaksonen, H., & Salmelin, R. (2012). Neural interactions at the core of phonological and semantic priming of written words. *Cerebral Cortex, 22*, 2305–2312. doi:10.1093/cercor/bhr307

Li, B., Daunizeau, J., Stephan, K. E., Penny, W., Hu, D., & Friston, K. (2011). Generalised filtering and stochastic DCM for fMRI. *NeuroImage, 58*, 442–457. doi:10.1016/j.neuroimage.2011.01.085

Li, B., Wang, X., Yao, S., Hu, D., & Friston, K. (2012). Task-dependent modulation of effective connectivity within the default mode network. *Frontiers in Psychology, 3*, 206. doi:10.3389/fpsyg.2012.00206

Londei, A., D'Ausilio, A., Basso, D., Sestieri, C., Gratta, C. D., Romani, G. L., et al. (2010). Sensory-motor brain network connectivity for speech comprehension. *Human Brain Mapping, 31*, 567–580. doi:10.1002/hbm.20888

Mclntosh, A. R., & Gonzalez-Lima, F. (1994). Structural equation modeling and its application to network analysis in functional brain imaging. *Human Brain Mapping, 2*, 2–22.

Mody, M., Wehner, D. T., & Ahlfors, S. P. (2008). Auditory word perception in sentence context in reading-disabled children. *Neuroreport, 19*, 1567–1571. doi:10.1097/WNR.0b013e328311ca04

Moran, R., Pinotsis, D. A., & Friston, K. (2013). Neural masses and fields in dynamic causal modeling. *Frontiers in Computational Neuroscience, 7*, 57. doi:10.3389/fncom.2013.00057

Nolte, G., Ziehe, A., Nikulin, V. V., Schlogl, A., Kramer, N., Brismar, T., et al. (2008). Robustly estimating the flow direction of information in complex physical systems. *Physical Review Letters, 100*, 234101. doi:10.1103/PhysRevLett.100.234101

Noppeney, U., Price, C. J., Penny, W. D., & Friston, K. J. (2006). Two distinct neural mechanisms for category-selective responses. *Cerebral Cortex, 16*, 437–445. doi:10.1093/cercor/bhi123

Perrone-Bertolotti, M., Kujala, J., Vidal, J. R., Hamame, C. M., Ossandon, T., Bertrand, O., et al. (2012). How silent is silent reading? Intracerebral evidence for top-down activation of temporal voice areas during reading. *The Journal of Neuroscience, 32*, 17554–17562. doi:10.1523/JNEUROSCI.2982-12.2012

Pinotsis, D. A., Brunet, N., Bastos, A., Bosman, C. A., Litvak, V., Fries, P., et al. (2014). Contrast gain control and horizontal interactions in V1: A DCM study. *NeuroImage, 92*, 143–155. doi:10.1016/j.neuroimage.2014.01.047

Pinotsis, D. A., & Friston, K. J. (2014). Gamma oscillations and neural field DCMs can reveal cortical excitability and microstructure. *AIMS Neuroscience Special Issue on Gamma Oscillations, 1*, 4–24.
Pinotsis, D. A., Moran, R. J., & Friston, K. J. (2012). Dynamic causal modeling with neural fields. *NeuroImage, 59*, 1261–1274. doi:10.1016/j.neuroimage.2011.08.020
Pinotsis, D. A., Schwarzkopf, D. S., Litvak, V., Rees, G., Barnes, G., & Friston, K. J. (2013). Dynamic causal modelling of lateral interactions in the visual cortex. *NeuroImage, 66*, 563–576. doi:10.1016/j.neuroimage.2012.10.078
Saur, D., Schelter, B., Schnell, S., Kratochvil, D., Küpper, H., Kellmeyer, P., et al. (2010). Combining functional and anatomical connectivity reveals brain networks for auditory language comprehension. *NeuroImage, 49*, 3187–3197. doi:10.1016/j.neuroimage.2009.11.009
Schreiber, T. (2000). Measuring information transfer. *Physical Review Letters, 85*, 461–464. doi:10.1103/PhysRevLett.85.461
Seghier, M. L., Bagdasaryan, J., Jung, D. E., & Price, C. J. (2014). The importance of premotor cortex for supporting speech production after left capsular-putaminal damage. *The Journal of Neuroscience, 34*, 14338–14348. doi:10.1523/JNEUROSCI.1954-14.2014
Seth, A. K., Barrett, A. B., & Barnett, L. (2015). Granger causality analysis in neuroscience and neuroimaging. *The Journal of Neuroscience, 35*, 3293–3297. doi:10.1523/JNEUROSCI.4399-14.2015
Stephan, K. E., Penny, W. D., Moran, R. J., den Ouden, H. E., Daunizeau, J., & Friston, K. J. (2010). Ten simple rules for dynamic causal modeling. *NeuroImage, 49*, 3099–3109. doi:10.1016/j.neuroimage.2009.11.015
Studdert-Kennedy, M., & Mody, M. (1995). Auditory temporal perception deficits in the reading-impaired: A critical review of the evidence. *Psychonomic Bulletin & Review, 2*:508–514. doi:10.3758/BF03210986
Vicente, R., Wibral, M., Lindner, M., & Pipa, G. (2011). Transfer entropy--a model-free measure of effective connectivity for the neurosciences. *Journal of Computational Neuroscience, 30*, 45–67. doi:10.1007/s10827-010-0262-3
Wehner, D. T., Ahlfors, S. P., & Mody, M. (2007). The influence of semantic processing on phonological decisions in children and adults: A magnetoencephalography (MEG) study. *Journal of Speech, Language, and Hearing Research, 50*, 716–731. doi:10.1044/1092-4388(2007/050)
Yang, J., & Shu, H. (2012). The role of the premotor cortex and the primary motor cortex in action verb comprehension: Evidence from Granger causality analysis. *Brain Research Bulletin, 88*, 460–466. doi:10.1016/j.brainresbull.2012.04.006

Chapter 9
A Role for the Cerebellum in Language and Related Cognitive and Affective Functions

Peter Mariën

9.1 Introduction

At the beginning of the twentieth century, several authors defined the role of the cerebellum as a modulator of motor functions including diadochokinesia, tonus, coordination, and motor speech production (Babinski 1902; Luciani 1891; Holmes 1922). Although from time to time clinical case descriptions and experimental evidence from animal studies dating back to the early part of the nineteenth century suggested an association between cerebellar pathology and a variety of nonmotor cognitive and affective dysfunctions, a causal relationship remained unexplored and was dismissed for several decades. During the past three decades converging evidence from a wealth of neuroanatomical, neuroimaging, and clinical studies has unambiguously demonstrated that the cerebellum is also involved in cognitive, affective and linguistic processing. Neuroanatomical studies revealed that the cerebellum is closely linked in a reciprocal way to the autonomic, limbic, and associative regions of the supratentorial cortex[1] (for a review, see Schmahmann 2004). In addition, cortical areas send information to the cerebellum via the basilar pons (Schmahmann and Pandya 1997), and deep cerebellar nuclei send information back to the cortical association areas through dentatothalamic pathways (Middleton and Strick 1997) (Fig. 9.1).

[1] Refers to all of the cerebral cortex lying above and anterior to the tentorium cerebelli. It is that part of the dura that 'tents' the cerebellum on its superior surface separating it from the inferior occipital cortex.

P. Mariën (✉)
Department of Clinical and Experimental Neurolinguistics, Vrije Universiteit Brussel, Brussels, Belgium

Department of Neurology and Memory Clinic, Ziekenhuis Netwerk Antwerpen – Middelheim, Antwerp, Belgium
e-mail: peter.marien5@telenet.be

M. Mody (ed.), *Neural Mechanisms of Language*, Innovations in Cognitive Neuroscience, DOI 10.1007/978-1-4939-7325-5_9

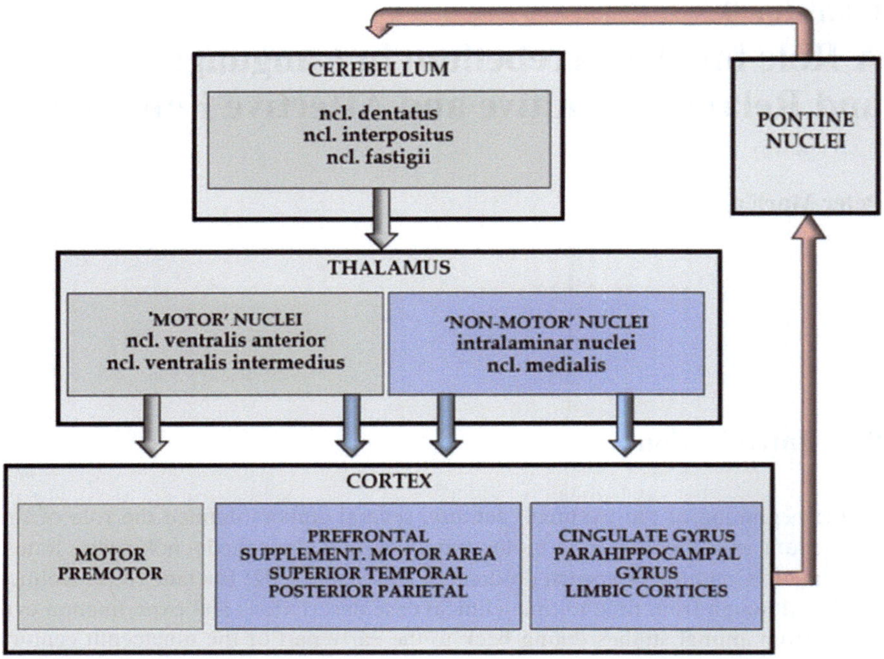

Fig. 9.1 Diagram depicting the cerebello-cerebral connectivity network underlying cognitive and affective processes. The feedback or efferent loop originates from the deep nuclei of the cerebellum that project to the motor (*grey*) and nonmotor (*blue*) nuclei of the thalamus. In turn, the motor nuclei of the thalamus (ncl. ventralis anterior and intermedius) project not only to motor and premotor cortices (*grey arrow*) but also to nonmotor areas among which are the prefrontal cortex, the supplementary motor area, the superior temporal and posterior parietal regions (*blue arrow*). The nonmotor nuclei of the thalamus project to the cingulate gyrus, the parahypocampal region, and the limbic cortices (*blue arrows*). The feedforward or afferent system of the cerebello-cerebral circuit is composed of corticopontine and pontocerebellar mossy fiber pathways (*red arrows*) (after Schmahmann and Pandya (1997) and from Mariën et al. (2013b)).

Studies with positron emission tomography (PET) in healthy subjects have provided evidence for cerebellar involvement in nonmotor language functions. Indeed, PET investigations of healthy subjects revealed a consistent and simultaneous activation of the supratentorial language areas (Broca's area) and the right cerebellar hemisphere during a semantic word association task (Petersen et al. 1988, 1989). In-depth neuropsychological investigations of an etiologically heterogeneous group of patients with focal and diffuse cerebellar lesions allowed clinicians to identify a variety of generally mild but clinically significant linguistic, cognitive, and affective deficits after cerebellar damage. This approach resulted in a large number of case reports describing cognitive and linguistic symptoms following isolated cerebellar lesions. Subsequently, many studies with a robust methodology including large cohorts of patients with cerebellar disorders and carefully matched healthy controls were performed to identify the multifaceted modulatory role of the cerebellum in a variety of nonmotor cognitive and affective functions.

In this chapter a concise overview of the modulating role of the cerebellum in language as well as in a variety of related cognitive and behavioral-affective processes is presented.

9.2 The Cerebellum and Language

9.2.1 Verbal Fluency and Lexical Retrieval

In the late 1980s, PET activation studies with healthy subjects provided the first evidence in support of the emerging view that the cerebellum might be implicated in linguistic processes (Leiner et al. 1986). In a PET experiment with healthy subjects, Petersen et al. (1988, 1989) showed that during the production of semantically related verbs in response to visually presented nouns, activation of dominant Broca's area and the contralateral cerebellar hemisphere occurred. For the first time a consistent pattern of activation was shown that was not due to motor verbal responses but to nonmotor linguistic processes subserving semantic word association processes. Notwithstanding variations on the original task design, subsequent studies in healthy subjects consistently reproduced activation of the right lateral cerebellum during word generation tasks (Raichle et al. 1994; Papathanassiou et al. 2000). Hubrich-Ungureanu et al. (2002) investigated the pattern of lateralized activations in a left and right-handed volunteer by means of functional magnetic resonance imaging (fMRI) during a silent verbal fluency task. In the right-handed subject with typical left hemisphere language dominance, regions of activation not only included the language dominant left fronto-parietal cortex but, as expected, also the contralateral right cerebellar hemisphere. In the left-handed subject with atypical right hemisphere language dominance a reversed pattern of language activation was found, reflected by crossed cerebral-cerebellar activations involving the right cerebral and the left cerebellar hemisphere. The study concluded that cerebellar involvement in language processing is contralateral to the activation of the cerebral cortex, even under conditions of different language dominance. As demonstrated in Fig. 9.2a, b, this crosswise functional cerebello-cerebral network subserving lexical retrieval processes was found in a right-handed patient with atypical cerebral language dominance. Atypical activations of Broca's homologue in the right hemisphere were accompanied by contralateral activations in the left cerebellar hemisphere during an fMRI noun to verb association task.

The role of the cerebellum in phonemic and semantic fluency tasks has recently also been investigated by means of Transcranial Magnetic Stimulation (TMS) using continuous theta burst stimulation (cTBS) (Arasanz et al. 2012). Twenty-seven healthy subjects were randomly assigned to one or two groups for application of cTBS to the posterior-lateral cerebellum, and the left or right cerebellar hemisphere. The subjects first participated in a phonological verbal fluency task (with letters F, A, S or P, R, W) followed by a semantic verbal fluency task consisting of the categories animals or groceries. Arasanz et al. (2012) hypothesized that the number of

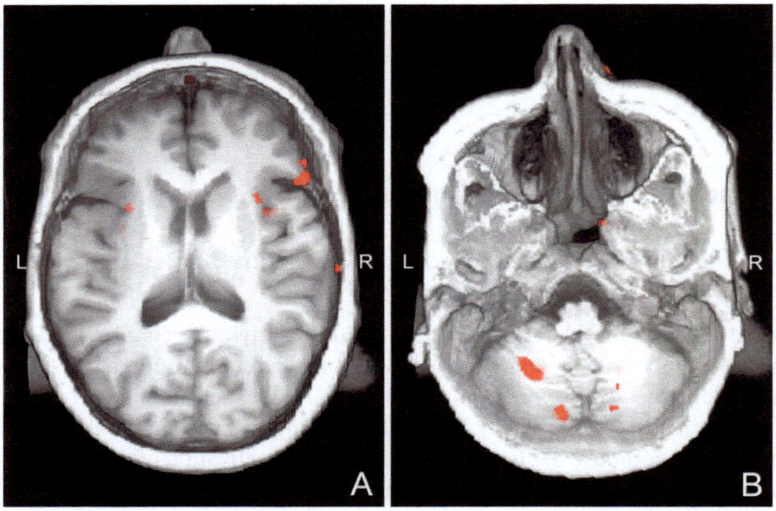

Fig. 9.2 fMRI images of the brain using a verb generation task disclosing predominantly activation in the right prefrontal and insular regions (**a**) associated with activation in the left cerebellar hemisphere (**b**)

category switches between the subcategories of words is a measure of mental flexibility, which is the highest during the first 15 s of the task. It was found that within the first 15 s of each trial, subjects with right cTBS had significantly lower switching scores after stimulation. In addition, the study also provided evidence in favor of the hypothesis that the cerebellum is crucially implicated in executive control via a dense network of cerebello-cerebral connections to the (pre)frontal cortex (Petersen et al. 1989; Raichle et al. 1994; Martin et al. 1995; Grabowski et al. 1996; Schlösser et al. 1998; Gourovitch et al. 2000; Papathanassiou et al. 2000; Hubrich-Ungureanu et al. 2002; Arasanz et al. 2012).

A number of clinical studies of patients with focal and diffuse cerebellar damage have confirmed the involvement of the cerebellum in word production processes. In an early study, Fiez et al. (1992) described a 41-year-old, right-handed lawyer who despite high-level conversational skills presented with semantic retrieval deficits after a vascular lesion of the right cerebellar hemisphere. Leggio et al. (1995, 2000) compared patients with focal and degenerative left and right cerebellar lesions with healthy control subjects using cluster analysis. They showed that right cerebellar damage particularly affects phonological fluency while sparing semantic fluency. However, a small number of subsequent studies disclosed no evidence of a lateralized impact as reduced verbal fluency was observed in patients with either left or right cerebellar lesions (Cook et al. 2004; Whelan and Murdoch 2005). These observations contrast with a recent study of Schweizer et al. (2010) who investigated 22 patients with chronic, unilateral cerebellar lesions (12 patients with left and ten patients with right cerebellar lesions). These authors demonstrated that the patient group with right cerebellar damage produced significantly fewer words in a phonemic fluency task in comparison to both the patient group with left cerebellar damage and the group of healthy control subjects. Performance characteristics of the right cerebellar lesion group were highly similar to the performance characteristics of patients with left prefrontal lesions. On the basis of these overt resemblances, Schweizer et al. (2010) suggested that the findings reflect a lateralized network effect consisting of a supratentorial, left prefrontal, and infratentorial right cerebellar system for the modulation of attention/executive function tasks.

9.2.2 Syntax Impairment

A number of studies have demonstrated that grammatical and syntactic disorders may result from focal cerebellar damage. Disruption of grammatical processing was for the first time described by Silveri et al. (1994) who found an association between focal vascular damage of the right cerebellum and transient expressive agrammatism, characterized by the omission of free-standing grammatical morphemes, the omission of auxiliaries and clitics, and substitutions of bound grammatical morphemes. Single-photon emission computed tomography (SPECT) scan of the brain showed a relative hypoperfusion affecting the entire left cerebral hemisphere, more stable and consistent in the left posterior temporal region. This patient's selective speech production impairment was, however, interpreted as a non-linguistic, "peripheral" disorder and it was hypothesized that agrammatism may be the result of the patient's adaptation to a deficit outside the mental linguistic system. In agreement with the view that the cerebellum acts as a controller and regulator of the temporal aspects of motor as well as nonmotor processes (timing hypothesis) this deficit was considered to be the result of a general timing disorder. Since then, several other patients have been reported who presented with expressive and/or receptive agrammatism following cerebellar damage (Mariën et al. 1996; Gasparini et al. 1999; Zettin et al. 1997). Strelnikov et al. (2006) investigated the brain mechanisms underlying prosodic segmentation and pitch processing in syntactically correct perception of phrases using PET. Twelve right-handed healthy subjects listened to phrases in which different prosodic segmentation substantially changed the meaning of the phrase. Activation was seen in the right dorsolateral prefrontal cortex and medial posterior area of the right cerebellum. According to the authors, the right posterior prefrontal cortex represents the functional overlap of brain networks of emotion, prosody, and syntax perception, whereas the right cerebellar activation was related to the assessment of time intervals necessary for different sensorimotor and cognitive activities (Ivry and Richardson 2002; Salman 2002), as in the estimation of phonetic and semantic borders of syntagmata, or to the maintenance of the phrase structure in working memory during processing (Mariën et al. 2001).

9.2.3 Aphasia

The notion of cerebellar-induced aphasia (Hassid 1995; Fabbro et al. 2000, 2004) emerged as a result of a co-occurrence of a spectrum of linguistic impairments affecting the phonological, lexico-semantic, and syntactic domains to different degrees after acute cerebellar damage. Mariën et al. (1996, 2000) described a 73-year-old, right-handed patient who presented with a dynamic aphasia-like language disorder after an ischemic lesion in the vascular territory of the right superior cerebellar artery. The patient's language disorder was characterized by a marked dissociation between nearly normal imposed (e.g., naming, repetition) and severely disrupted spontaneous language consisting of a severe lack of spontaneous speech initiation, and effortful and fragmented attempts to formulate ideas. In addition, there were word-finding difficulties in conversational speech, marked expressive and receptive agrammatism, and reading and writing deficits. Mariën et al. (1996, 2000) labelled their patient's language disorder as *cerebellar-induced aphasia*. In this patient, follow-up SPECT studies revealed a significant hypoperfusion in the right cerebellum and in the anatomoclinically suspected prefrontal language region of the left hemisphere. At follow-up, changes in perfusional patterns paralleled the alterations in the neurolinguistic profile. Aphasia-like phenomena following right cerebellar damage were considered to result from a loss of excitatory impulses through the cerebello-ponto-thalamo-cortical pathways (Mariën et al. 1996). In agreement with these findings, Mariën and coworkers subsequently reported an additional number of right-handed patients who presented aphasic symptoms in association with cognitive and behavioral problems after right cerebellar damage (Mariën et al. 2007, 2009; Baillieux et al. 2010; De Smet et al. 2012).

Karacı et al. (2008) evaluated in 20 patients with ischemic lesions of the cerebellum and 20 control subjects the effects of focal cerebellar damage on language functions and the relation between these functions and lesion type, age, and education level. A variety of aphasic symptoms were identified at the level of speech production, comprehension, repetition, naming, reading, and writing. However, with respect to lateralization (left vs. right) and vascular territory (posterior inferior cerebellar artery (PICA) vs. superior cerebellar artery (SCA)), no significant effects were found. Recently, Blancart et al. (2011) described an 83-year-old right-handed man who suffered from aphasia after a left cerebellar infarction. Dysarthria, anomia, agrammatism, comprehension deficits, reading and writing difficulties characterized this patient's speech and language. Two months post-stroke, anomia, and agrammatism still persisted while reading and writing abilities had improved and comprehension had nearly normalized. In addition, cognitive and behavioral-affective abnormalities were reported, including disorientation, apathy, stiff and obsessive behavior, aggressiveness, and daytime hypersomnia. PET studies demonstrated hypometabolism in the left cerebellar hemisphere and the bilateral temporoparietal regions. Based on these findings, Blancart et al. (2011) suggested that the left cerebellar infarction was responsible for the language deficits in the acute phase and that the cerebellar lesion played a major precipitating role in the development of cognitive and behavioral problems two months post-onset. Although the authors mention the concept of crossed aphasia following left cerebellar injuries, no information is provided with regard to supratentorial language dominance nor did they elaborate on the mechanism of ipsilateral cortical diaschisis[2], a phenomenon observed following focal cerebellar damage (De Smet and Mariën 2012).

9.2.4 Alexia and Dyslexia

Although reading impairment may follow cerebellar damage, only a handful of patients have been reported in the literature. Moretti et al. (2002a) investigated the impact of cerebellar lesions on reading skills in ten patients with cerebellar vermis/paravermis lesions compared to ten right-handed controls. The patient group demonstrated a lower degree of accuracy in reading words and sentences. They made errors both at letter and word level. The authors suggested that acquired dyslexia in patients with cerebellar damage may be related either to imperfect oculomotor control (nystagmus), or to disruption of the cerebellar-encephalic projections connecting the cerebellum to the supratentorial areas implicated in language as well as in attentional and alerting processes.

Mariën et al. (2009) reported a patient who after an ischemic stroke in the vascular territory of the right superior cerebellar artery (SCA) presented with the cerebellar cognitive-affective syndrome (CCAS) associated with visual dyslexia and surface dysgraphia. Acute phase data revealed a generalized cognitive decline and mild transcortical sensory aphasia. In the lesion phase of the stroke (i.e., 3 weeks to 4 months poststroke), neurobehavioral abnormalities mainly comprised executive dysfunctions, disrupted divided attention, deficient visual–spatial organization and a range of behavioral abnormalities. In-depth neurolinguistic investigations of reading and writing skills were consistent with a diagnosis of visual dyslexia and surface dysgraphia. Reading of words and performance on visual lexical decision tasks involving words and nonwords was severely disrupted and predominantly characterized by visual errors. In addition, writing irregular and ambiguous words resulted in regularization errors (phonologically plausible errors based on phoneme-grapheme correspondence rules). In the absence of any structural damage in the supratentorial brain regions, a quantified SPECT study showed a relative hypoperfusion in the right cerebellar hemisphere and the left medial frontal lobe (Fig. 9.3).

Mariën et al. (2009) hypothesized that the cognitive and linguistic deficits result from functional disruption of the cerebellar-encephalic pathways, connecting the cerebellum to the frontal supratentorial areas which subserve attentional and planning processes. Functional disruption of the anatomoclinically suspected brain regions was reflected on SPECT by the phenomenon of crossed cerebello–cerebral diaschisis.

[2]Diaschisis, a concept introduced at the beginning of the twentieth century by Constantin Von Monakow, stands for the distant functional impact of a brain lesion on an anatomically connected and structurally intact brain region. Depressed function of this intact region is considered to result from a decrease or loss of excitatory impulses from the anatomically connected, lesioned area.

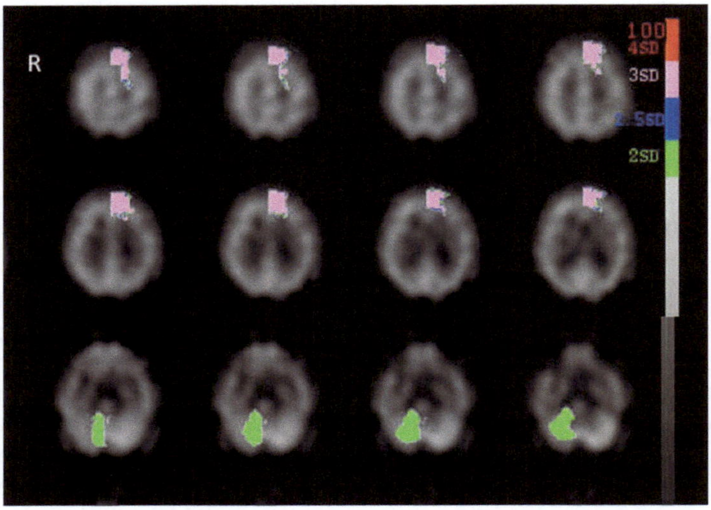

Fig. 9.3 Quantified ECD-SPECT scan 5 weeks after a right cerebellar stroke shows a hypoperfusion in the right cerebellar hemisphere and the left medial frontal area (crossed cerebello-cerebral diaschisis)

Possible involvement of the cerebellum in the pathogenesis of dyslexia has recently been postulated as well in a large number of studies (Nicolson et al. 2001; Finch et al. 2002; Rae et al. 2002; Eckert et al. 2003; Pernet et al. 2009). Structural MRI studies conducted in adults with dyslexia have demonstrated cerebellar anomalies (Brown et al. 2001; Rae et al. 2002). Fawcett and Nicolson (1999) studied 59 dyslexic children and 67 matched control subjects and showed subtle cerebellar-related abnormalities in the dyslexic group such as difficulties in motor skills, automatization, information processing, speed, and balance. On the basis of these "soft neurological signs," Nicolson et al. (1995, 1999, 2001) introduced the "cerebellar deficit hypothesis" to explain dyslexia. According to this hypothesis, the automatization of learned skills such as articulation, reading, spelling, and phonological abilities is disrupted as a result of a cerebellar dysfunction (Fawcett and Nicolson 1999; Nicolson et al. 1995). A cerebellar deficit in young children might induce a delay in the automatization process of articulation, causing deficits in phonological awareness. Therefore, cerebellar maturational impairments might result in a "phonological core deficit," which provides an explanatory framework for various aspects of developmental dyslexia (Nicolson et al. 1999). Evidence to support this hypothesis was provided by a PET study in six dyslexic adults versus an age-matched control group of six healthy subjects (Nicolson et al. 1999) who performed either an automatic prelearned sequence or a novel sequence of finger movements. In the group of the dyslexics, significantly lower brain activations were found in the right cerebellar cortex and the left cingulate gyrus when executing the prelearned sequence and in the right cerebellar cortex when learning the new sequence. Baillieux et al. (2009) investigated 15 dyslexic children and seven carefully matched (age, gender, IQ) control subjects by means of functional neuroimaging (fMRI) using a noun-verb association paradigm. Comparison of activation patterns between dyslexic and control subjects revealed distinct and significant differences in activation patterns at both the cerebral and cerebellar level. Control subjects showed well-defined and focal activation patterns bilaterally distributed in the frontal and parietal lobes and the posterior regions of the cerebellar hemispheres. The dyslexic children, however, presented widespread and significantly more diffuse activations at the cerebral and cerebellar levels. Cerebral activations were observed in frontal, parietal, temporal, and occipital areas. Activations in the cerebellum were found predominantly in the cerebellar cortex, including Crus I, Crus II, hemispheric lobule VI, VII and vermal lobules I, II, III, IV, and VII (Fig. 9.4). Given the widespread activation in the cerebellum in the dyslexic group the authors suspected a defect of the intra-cerebellar distribution of activity, suggesting a disorder of the processing or transfer of information within the cerebellar cortex.

Nicolson and Fawcett (2011) published a review on the role of the cerebellum in various developmental disorders such as dyslexia, dysgraphia, and procedural learning and suggested that according to the cerebellar deficit hypothesis only the language-related cerebellar regions, including lobules VI and VIIB, are involved in dyslexia. Other regions in the cerebellum may be affected as well. According to their neural systems framework, dyslexia is associated with the language-based component (Broca's area and the right lateral cerebellum), whereas dysgraphia is

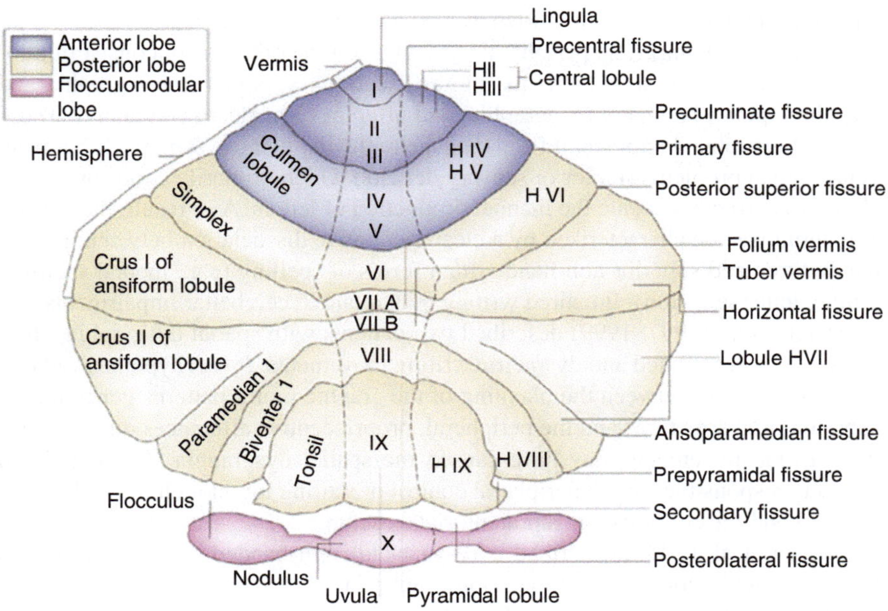

Fig. 9.4 Cerebellar cortical anatomy: view of the cerebellar cortex presenting the lobes and lobuli by name and number (after Manni and Petrosini 2004)

primarily related to the motor component (the motor cortex and the cerebellum). The neural systems framework is derived from Ullman's (2004) model of a procedural learning motor system for motor skills such as handwriting and a procedural learning system for language skills and habits. This model includes the frontal cortex, the parietal cortex, the superior temporal cortex, and some subcortical structures such as the basal ganglia and the cerebellum. Consequently, Nicolson and Fawcett (2011) suggested that developmental dyslexia arises from impaired performance of the "procedural learning system for language," which comprises the prefrontal cortex (around Broca's area), the parietal cortex, the basal ganglia and the cerebellum. It is possible that dyslexic children also have difficulties with the motor procedural learning system but these seem not necessary to induce dyslexia.

9.2.5 Agraphia

Agraphia is a generic term denoting various types of writing disorders that result from acquired neurological damage. On the basis of their semiological characteristics, agraphic phenomena can be classified as either of the central (linguistic) or the peripheral (non-linguistic) type. The central agraphias comprise lexical (or surface) agraphia, phonological agraphia, deep agraphia, semantic agraphia and agraphia due to impairment of the graphemic buffer and they involve disruption of the linguistic system: they are characterized by qualitatively similar spelling errors across all output modalities (e.g., in written as well as in oral spelling, typing, letter selection, and sequencing). The peripheral agraphias, on the other hand, consist of allographic agraphia, apraxic agraphia, motor execution agraphia (micrographia and megalographia), hemianoptic agraphia, and afferent or neglect dysgraphia). These forms of agraphia do not result from damage to the linguistic system itself but from neurological problems (motor or sensory deficits) which primarily compromise the ability to correctly execute the manual production of letters. As a result, the peripheral agraphias are characterized by a clear qualitative dissociation between inferior handwritten and superior non-handwritten forms of spelling (e.g., mental spelling, typing, letter selection). Impaired writing may relate to cerebellar impairments.

Silveri et al. (1997, 1999) described two patients with spatial dysgraphia, characterized by segmented and dysmetric writing movements. It was hypothesized that a discoordination between the planning of the graphic motor patterns generated by supratentorial structures and the peripheral, proprioceptive afferences during ongoing writing movements may have caused the spatial dysgraphia. The functional pathway responsible for the peripheral control of writing might include the left cerebellum and the contralateral supratentorial structures.

Mariën et al. (2007) described a 72-year-old right-handed civil engineer who presented with apraxic agraphia, mild aphasia, dysexecutive symptoms, and behavioral and affective changes after a hemorrhagic lesion affecting the right cerebellar hemisphere. At one-year follow-up, apraxic agraphia, executive dysfunctions, and behavioral changes persisted (Fig. 9.5).

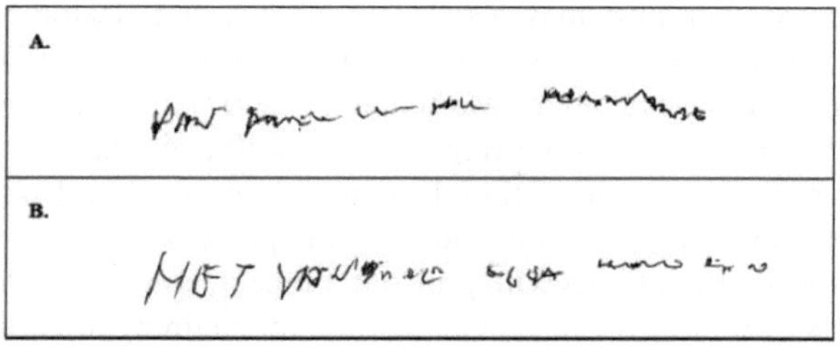

Fig. 9.5 Handwriting sample demonstrating the characteristic features of apraxic agraphia including poor and irregular letter formation, spatial distortions, stroke omissions resulting in incomplete letter forms, redundant insertions of anomalous strokes, and illegible scrawls: writing of lower (1**a**) and upper (1**b**) case words to dictation at 1 month

Quantified SPECT studies at 1 and 6 months post-stroke revealed perfusion deficits in the right cerebellar hemisphere as well as in the medial and lateral regions of the left prefrontal hemisphere. De Smet et al. (2012) reported three additional cases with apraxic agraphia following vascular damage to the cerebellum. The first patient presented with dysarthria, disrupted language dynamics, mild comprehension difficulties, reduced verbal fluency, and apraxic agraphia. Deficits in memory, attention, visuo-spatial planning, and executive functions were found as well. The patient's neurolinguistic deficits were consistent with a diagnosis of dynamic aphasia associated with anomia and apraxic agraphia. The second patient had ataxic dysarthria, apraxic agraphia, and mild attention problems. The third patient presented with disturbed frontal problem-solving, impaired mental flexibility, and apraxic agraphia. Although the patient occasionally had difficulties recalling the shapes of some graphemes, letter recognition was entirely normal. No difficulties were found matching lower- and upper-case letters and vice versa. Given the absence of sensorimotor disturbances affecting the writing limb acute vascular ischemic damage to the cerebellum could be held responsible in all three cases for the writing disturbance matching a diagnosis of apraxic agraphia.

Apraxic agraphia reflects damage to processing components involved in the programming of skilled movements for writing. Mechanisms responsible for this disruption include the destruction or disconnection of stored graphic motor engrams or damage to systems associated with translating the information on graphic motor engrams into graphic innervatory patterns to specific muscles (Rapcsak and Beeson 2000). Graphic motor engrams are stored in the parietal lobe, whereas the frontal premotor areas (dorsolateral premotor cortex and the supplementary motor area) are involved in translating these programs into graphic innervatory patterns (Rapcsak and Beeson 2000). Consequently, parietal damage results in the destruction of stored spatiotemporal representations for writing movements and frontal premotor lesions interfere with the execution of appropriate motor commands to specific muscle systems. However, De Smet et al.'s (2012) survey clearly revealed that no cases have been reported in whom apraxic agraphia resulted from isolated vascular lesions restricted to the prefrontal lobe. In addition, analysis of the study corpus consisting of 25 vascular cases reported since the first description by Heilman et al. (1973)

showed that apraxic agraphia may also occur after various lesion locations (for a review see De Smet et al. 2012). Given structural neuroimaging evidence of unaffected prefrontal and parietal areas, Mariën et al. (2007) and De Smet et al. (2012) hypothesized that their patients' writing deficits result from damage to the cerebellar–encephalic projections, connecting the cerebellum to the prefrontal supratentorial areas which subserve attentional and planning processes (Moretti et al. 2002b). Clinical observations of patients with prefrontal lobe dysfunctions also suggest that complex aspects of writing such as planning and maintaining attention may be disturbed (Ardila and Surloff 2006). This view of a functional disruption of the prefrontal brain regions is supported by SPECT findings which revealed the phenomenon of cerebello–cerebral diaschisis, reflecting the functional impact of the cerebellar lesion on a distant supratentorial region crucially involved in the execution of written language, due to a lack of excitatory impulses (Baron et al. 1981; Marien et al. 2001). A similar pattern of decreased perfusion in the anatomoclinically suspected prefrontal and cerebellar brain regions crucially involved in the planning and execution of skilled motor actions was recently identified in a 15-year-old left-handed patient with apraxic dysgraphia (Mariën et al. 2013a). It was hypothesized that in the absence of structural brain damage, disrupted development of handwriting skills in this patient might reflect incomplete maturation of the cerebello-cerebral network involved in planned skilled actions (Mariën et al. 2013a).

A few functional neuroimaging studies with fMRI have shown involvement of the right cerebellar hemisphere in writing beyond the pure motor control level (Katanoda et al. 2001; Longcamp et al. 2003). Beeson et al. (2003) conducted an fMRI study on the neural substrates of writing and main cerebral activation was seen in the superior part of the left parietal lobe and left inferior and middle frontal gyri. Their study confirmed the role of the superior parietal and frontal premotor areas in translating orthographic information into appropriate hand movements.

9.2.6 Metalinguistic Skills

In a number of studies Murdoch and coworkers investigated higher-level language skills and metalinguistic abilities in patients with cerebellar lesions (Cook et al. 2004; Whelan and Murdoch 2005; Murdoch and Whelan 2007). Murdoch and Whelan (2007), for instance, described ten patients with left primary cerebellar strokes who experienced difficulties with definitions and multiple definitions tests, with figurative and ambiguous language tests, with word association tasks, with antonym/synonym generation, and with the interpretation of semantic absurdities. These observations support the hypothesis that left cerebellar damage may disrupt high-level language skills. In addition, Murdoch and coworkers argued that tasks involving the manipulation of novel situations and lexico-semantic operations, as well as the development of language and monitoring strategies require frontal lobe support in their manipulation (Copland et al. 2000). Although frontal lobe involvement may explain the presence of language deficits after left cerebellar damage, the authors alternatively suggested that left cerebellar lesions may induce language deficits typically associated with right cerebral damage (Kempler et al. 1999) via the phenomenon of crossed cerebello-cerebral diaschisis. Although no functional neuroimaging studies were conducted in this study, Murdoch and Whelan (2007) concluded that the cerebellum is involved in the process of refining and modulating language functions presumably via excitatory impulses reaching the prefrontal cortex from the cerebellum via the basal ganglia and the thalamus.

9.2.7 *Summary*

In addition to its long-established crucial role in coordinating motor speech production, clinical and experimental studies with patients suffering from etiologically different cerebellar disorders have identified involvement of the cerebellum in a variety of nonmotor language functions as well. A wealth of studies has demonstrated that the cerebellum is crucially implicated in various aspects of linguistic processing, including motor speech planning, language dynamics and verbal fluency, phonological and semantic word retrieval, expressive and receptive syntax processing, various aspects of reading and writing and even aphasia-like phenomena resembling dynamic and transcortical motor aphasia.

9.3 Cerebellar-Induced Syndromes

9.3.1 *The Cerebellar Cognitive Affective Syndrome (CCAS)/ Schmahmann's Syndrome*

Although from time to time, descriptions of clinical cases and experimental evidence from animal studies dating back to the early part of the nineteenth century already suggested an association between cerebellar disorders and nonmotor cognitive and affective dysfunctions, a causal relationship remained unexplored for several decades. In the mid-1900s investigators started to examine a possible causal relationship between the cerebellum and cognition and emotion exemplified by the work of Snider, Dow, Heath, Cooper, and others (see Schmahmann 2010 for a review). This line of research laid a robust foundation for the rediscovery of the concept by Leiner and colleagues (e.g., 1989) who hypothesized that more recently evolved parts of the cerebellum contribute to learning, cognition and language, and by Schmahmann and colleagues (see Schmahmann 2010 for a review) who introduced the influential dysmetria of thought hypothesis; the latter provides an historical, clinical, neuroanatomical, and theoretical framework within which a cerebellar role in higher cognitive and affective processes could be considered. In this theory, the cerebellum is considered to act as a modulator of behavior function, maintaining it around a homeostatic baseline appropriate to the context. In the way the cerebellum regulates motor function (rate, force, rhythm, and accuracy of movements), it regulates the speed, capacity, consistency, and appropriateness of affective and cognitive processes.

In the late 1990s, Schmahmann and Sherman (1998) introduced in a much cited paper the concept of cerebellar cognitive affective syndrome (Schmahmann's Syndrome; Manto and Mariën, 2015) to identify a range of cognitive and affective disturbances in patients with isolated cerebellar lesions. Based upon bedside screening and formal neuropsychological testing of 20 patients, Schmahmann and Sherman (1998) identified a range of cognitive, linguistic, and affective symptoms following cerebellar damage. The core features of this syndrome consist of (1) executive dysfunctions such as disturbances in planning, set-shifting, abstract reasoning, and working memory, (2) visuo-spatial deficits, such as impaired visuo-spatial organization and memory, (3) mild language symptoms including agrammatism and anomia, and (4) behavioral-affective disturbances, consisting of blunting of affect or disinhibited and inappropriate behavior. Anatomoclinical analysis revealed that lesions of the posterior lobe of the cerebellum (PICA territory) resulted in cognitive symptoms, while the vermis was consistently damaged in patients with behavioral-affective disturbances (Schmahmann and Sherman 1998). In contrast to subjects with SCA lesions, Exner et al. (2004) found a consistent pattern of memory impairment, executive disturbances and emotional withdrawal in patients with infarcts in the PICA territory. However, Neau et al. (2000) did not find any differences between the cognitive consequences of infarctions in the PICA or the SCA territory. Furthermore, patients with SCA lesions have been reported with clinically significant cognitive or linguistic disturbances (Mariën et al. 2001, 2009). Baillieux et al. (2010) investigated 18 adult patients with isolated cerebellar damage of whom 15 (83%) presented with cognitive impairments and/or behavioral-affective disturbances. Analysis of the neuropsychological profiles revealed a clear tendency of functional lateralization within the cerebellum: left cerebellar damage was related to typical non-dominant, right-hemisphere dysfunctions, such as attention deficits and visuo-spatial disturbances, while right cerebellar damage was associated with typical dominant, left-hemisphere deficits, such as disrupted language skills. There were no significant differences between SCA and PICA lesions. In addition, functional neuroimaging studies by means of quantified ECD SPECT demonstrated an association between supratentorial hypoperfusion and the observed neuropsychological deficits: Seven out of eight patients with frontal hypoperfusion presented with associated neuropsychological deficits, including executive dysfunction and/or behavioral disturbances.

Tedesco et al. (2011) investigated the expression of CCAS with respect to vascular lesion topography and the involvement of the deep cerebellar nuclei. Contrary to Baillieux et al. (2010), these authors did not find a lateralization effect but an effect of lesion distribution according to the vascular territory involved. Patients with PICA lesions performed significantly worse than patients with SCA lesions on tasks assessing verbal memory, language, visuo-spatial abilities. In addition, patients with lesions of the deep cerebellar nuclei had statistically significant lower scores for visuo-spatial memory, executive functions, attention, visuo-spatial, and sequencing skills. Although many studies have demonstrated a large spectrum of cognitive deficits following focal cerebellar damage, Alexander et al. (2012) only found minimal impairments in the chronic phase. Patients with right cerebellar lesions performed significantly worse on verbal fluency tasks and response control on the Stroop task in comparison to patients with left cerebellar damage and controls. It was suggested that clinically significant impairments in patients with focal cerebellar lesions are usually transient or mild. Their findings provide support for the hypothesis that lateralized cerebellar lesions may cause impairments in a parallel manner to contralateral prefrontal lesions.

Although clinical case descriptions dating back to the early part of the nineteenth century from time to time suggested an association between congenital cerebellar pathology and a variety of nonmotor cognitive as well as affective dysfunctions, a possible correlation was dismissed for decades. Steinlin et al. (2003) were among the first who reported in a study group of 11 adult patients with pure non-progressive congenital ataxia (with and without cerebellar hypoplasia) a consistent association between a number of cognitive and affective disturbances consistent with Schmahmann's Syndrome and congenital pathology of the cerebellum. Chheda et al. (2002) found a significant correlation between severity of motor, cognitive and affective deficits, and the extent of agenesis in their group of patients consisting of six children and two adults. CCAS in this group was characterized by executive dysfunction, visuo-spatial impairment, behavioral abnormalities, marked prosodic difficulties, and expressive language disturbances affecting two cases. CCAS was also found in a number of genetic conditions primarily affecting the cerebellum both structurally and functionally such as Gillespie syndrome (Mariën et al. 2008) and Joubert syndrome (Tavano et al. 2007). Tavano and Borgatti (2010) confirmed the presence of CCAS in a group of children and adults with different types of congenital malformative lesions of the cerebellum but observed a wide variability of cognitive and affective dysfunctions indicating different subtypes of CCAS.

From an anatomical point of view there is still no consensus regarding the anatomical parts of the cerebellum that subserve cognitive modulation. However, the symptoms observed in Schmahmann's Syndrome are consistent with predictions derived from anatomical and neuroimaging studies, which show extensive neural circuits connecting the prefrontal, temporal, posterior parietal, and limbic cortices with the cerebellum (Desmond 2001). According to Schmahmann (2004), these anatomical circuits constitute the structural basis for functional subunits, reflecting a topographic organization of motor, cognitive, and affective processing in the cerebellum, in which the anterior cerebellar lobe is mainly involved in motor functions, the posterior parts of the cerebellum in higher cognitive modulation, and the posterior vermis in affective processing (Stoodley and Schmahmann 2010; Stoodley et al. 2012). However, several studies and case reports demonstrate that there may be substantial variability regarding the functional anatomy of the cerebellum (Neau et al. 2000).

9.3.2 The Posterior Fossa Syndrome (PFS)

The posterior fossa syndrome (PFS), which may develop following acute cerebellar damage, is characterized by a broad range of linguistic, cognitive, and behavioral-affective disturbances (Pollack 1997). PFS may be considered as an aetiologically heterogeneous condition affecting both children and adults, but it most often occurs in paediatric patients after cerebellar tumor surgery. Although PFS has been documented in more than 350 cases, it is quite rare in adults (approximately 25 cases). In addition, PFS associated with vascular aetiologies is only reported in a very limited number of three adult cases (for a review see Mariën et al. 2013b). De Smet and Mariën (2012) described an adult patient with PFS following surgical evacuation of an intracerebellar hematoma. After 45 days of akinetic mutism, the patient's cognitive and behavioral profile closely resembled CCAS, characterized by visuo-spatial and attentional deficits, impaired frontal planning and problem solving, memory problems, reduced verbal fluency, decreased language dynamics and frontal-like behavioral problems such as apathy, behavioral and verbal inhibition, loss of facial expressions, and a withdrawn attitude. The authors suggested that in this patient, post-mutism cognitive and affective symptoms were related to the perfusional deficits in the anatomoclinically suspected prefrontal and right temporal cortical areas which subserve executive processing, behavioral–affective processes and spatial cognition.

Mariën et al. (2013c) reported longitudinal neuropsychological follow-up findings and pre- and postoperative SPECT in an adult patient with cognitive, behavioral, and affective symptoms before and after resection of an ependymoma in the posterior fossa This is the first patient in whom the phenomenon of pathological laughing and crying (PLC) was observed in the context of PFS, and the case provides evidence for the recently acknowledged role of the cerebellum in the contextual regulation of emotions and affect. During the phase of akinetic mutism, aggravation and marked extension of the perfusional deficits in the prefrontal brain regions were found. Mariën et al. (2013c) hypothesize that the phenomenon of cerebello-cerebral diaschisis in this patient suggests that PFS results from decreased transmission of excitatory impulses from the deep nuclei of the cerebellum through the dentatothalamic connections to the cortical areas crucially involved in cognition and behavioral and affective regulation (Mariën et al. 2001, 2003, 2009; Catsman-Berrevoets and Aarsen 2010; Miller et al. 2010).

As evidenced by a close parallels between SPECT and clinical findings, CCAS as well as PFS seem to reflect functional disruption of the cerebello-cerebral network involved in cognitive, behavioral, and affective functions. These findings may indicate that both syndromes share overt semiological features and a common pathophysiological substrate. Consequently, CCAS and PFS may both be regarded as cerebellar-induced clinical conditions showing different aspects of a spectrum that range in degree of severity and symptom duration.

9.4 Mechanisms of Cerebellar Involvement in Cognitive and Linguistic Processing

Several hypotheses have been advanced to explain the role of the cerebellum in various cognitive and linguistic processes such as non-motor associative learning, working memory, visuo-spatial abilities, verbal fluency, syntax, reading, and writing. The phenomenon of cerebello-cerebral diaschisis has often been suggested as a possible functional explanation of the cognitive, linguistic, and affective deficits in patients with cerebellar lesions (Mariën et al. 1996). Cerebello-cerebral diaschisis reflects the metabolic impact of a cerebellar lesion on a distant, but anatomically and functionally connected intact supratentorial region. Cortical regions crucially involved in cognitive and affective processing might become functionally suppressed because cerebellar damage induces a decrease or a loss of transmission of excitatory impulses from the deep nuclei of the cerebellum via the cerebello-ponto-thalamo-cerebral pathways to the supratentorial brain regions (Mariën et al. 2001). Numerous studies have convincingly demonstrated the crosswise functional impact of focal cerebellar damage on distant supratentorial regions that subserve cognitive processes and they contributed to the view of a functionally lateralized and topographic organization of the "cognitive cerebellum" (Botez-Marquard et al. 1994; Gottwald et al. 2004). However, the few reports of language deficits after left cerebellar lesions suggest that the correlation between the type of language disorders and cerebellar lateralization of linguistic functions may not be absolute (Cook et al. 2004; Fabbro et al. 2004; Murdoch 2010).

Another explanation for the involvement of the cerebellum in spatial function, language, verbal memory, and sequence processing is the sequencing hypothesis (Molinari et al. 2008). Evidence in support of this hypothesis is provided by animal models (Leggio et al. 1999), clinical (Silveri et al. 1994; Molinari et al. 2004), and functional neuroimaging studies (Doyon et al. 2003). This theory emphasizes the importance of the cerebellum in detecting patterns of incoming stimuli (temporal and spatial information) or in central circuit activities (Molinari et al. 2008). In order to accomplish the task of comparing previous and ongoing stimuli, data must be maintained in a working memory buffer. Cerebellar sequence processing should be considered within the network of cerebello-cortical connections. Consequently, damage to the cerebellum, depending on the cerebello-cerebral loop involved, may provoke different functional impairments such as defective processing of sensory stimuli (Leggio et al. 2011). Leggio et al. (2008) investigated the sequencing hypothesis across verbal, spatial, and behavioral domains in patients with focal or atrophic cerebellar damage. The authors administered a set of tests involving cartoon-like drawings to differentiate between verbal, spatial, and behavioral sequencing and found that patients with cerebellar damage had lower scores than control subjects irrespective of the material processed. When comparing right versus left cerebellar damage, patients with right cerebellar lesions obtained lower scores on tests requiring verbal processing, whereas patients with left cerebellar damage had lower scores on tests requiring the processing of non-verbal behavioral stimuli.

Another theory postulates that the cerebellum significantly contributes to the prediction of feedback or outcomes associated with sensory input or actions (Bellebaum and Daum 2011). The cerebellum provides internal models which need to be continuously modified and updated, based on the results of the comparison of their output with the output of the "controlled object" (real or imagined situations) (Ito 2008). Thus, if the predictions of the internal model do not accurately match reality an error signal is generated. Consequently, errors in predictions may result in deficits in error processing and error correction. Timmann et al. (2002) and Richter et al. (2004) demonstrated that patients with cerebellar damage may be impaired in associative learning tasks which might be due to an inability to update the internal model based on error feedback (Bellebaum and Daum 2011). According to this mechanism, the cerebellum does not only play an important role in the generation of predictions based on sensory stimuli but also in the generation of temporally accurate predictions. Evidence in support of this hypothesis was presented by clinical data of patients with cerebellar lesions who were impaired in the judgment of the duration of an auditory stimulus and the velocity of a moving visual stimulus (Ivry and Keele 1989; Ivry and Diener 1991). Patients with cerebellar lesions may also experience severe distortions during duration-discrimination tasks, suggesting a critical role of the cerebellum in the representation of temporal information (Hetherington et al. 2000; Spencer et al. 2007). This mechanism is also suggested to be involved in verbal working memory. According to Desmond et al. (1997), predictive control of the articulatory control process is necessary in order to update the phonological store dynamically. Consequently, prediction and updating are essential concepts in cerebellar processing of verbal working memory. Marvel and Desmond (2010) investigated cerebellar activity during the encoding, maintenance, and retrieval phases using a verbal working memory task and tried to find out whether cerebro-cerebellar activity is associated with the prediction of successful performance on a trial-by-trial basis. The authors found that the supplementary motor area and the dorsal cerebellar dentate are involved in encoding, and that the pre-supplementary area and the ventral dentate circuit are involved in retrieval. In addition, activity during the maintenance phase in the prefrontal lobe and the ventral dentate predicted subsequent accuracy of response to the probe during the retrieval phase. As a result, the study data consistently showed that the cerebro-cerebellar pathway is involved in accuracy prediction of successful performance.

9.5 Conclusion

The involvement of the cerebellum in cognitive, linguistic, and affective modulation has been overlooked for a very long time, due to its prominent role in motor functioning (Beaton and Mariën 2010). Although substantial progress has been made in understanding the functional role of the cerebellum in language and cognition, the precise role of the cerebellum in neurocognitive processing is not clear yet. Several pathophysiological and cognitive neuropsychological mechanisms have been suggested to explain various cognitive and linguistic deficits in patients with cerebellar lesions, including the phenomenon of cerebello-cerebral diaschisis, the sequencing hypothesis, and the role of the cerebellum in generating predictions. However, all hypotheses need further investigation to allow more consistent and firmer conclusions to be drawn about the exact nature of cerebellar computation. In addition, the question of a lateralized cerebellar involvement in cognitive modulation remains to be clarified. Although it has been demonstrated that specific cerebellar subsystems are involved in motor, cognitive, and affective processing, a better understanding of the functional topography of the cerebellum may clarify the contradictory findings with respect to neurobehavioral structure-function correlations.

Another area of interest in this relatively new research area relates to the prognosis of cognitive deficits following cerebellar damage. At present little is known about the long-term cognitive outcome. Richter et al. (2004) investigated the longitudinal outcome in 21 patients with cerebellar lesions with an average of 46 months following the onset of a cerebellar stroke. Their results indicated full recovery of cognitive disorders, except for marked impairments in verbal fluency. Schweizer et al. (2008) described a patient with a severe dysexecutive syndrome after a cerebellar arteriovenous malformation rupture. After 1 year of intensive rehabilitation, the patient's executive deficits had completely resolved. However, other studies did not confirm a positive prognosis following cerebellar lesions. De Smet et al. (2013) reported persistent linguistic, cognitive, and behavioral deficits in five children following posterior fossa tumor resection. Similar observations of persistent cognitive deficits were described by Neau et al. (2000), Fabbro et al. (2004), and Baillieux et al. (2006).

Findings suggest that classical language and neuropsychological tests may fail to detect subtle but significant cognitive changes after cerebellar damage (Mariën et al. 2000; Aarsen et al. 2004). Consequently, there is a need to develop more sensitive neuropsychological tools to identify the wide range of subtle neurocognitive repercussions after cerebellar dysfunction.

Refinement of insights into the functional role of the cerebellum in cognition and affect may also be accomplished by means of a close cooperation between the clinical neurosciences (neurology, neurolinguistics, neuropsychology), structural and functional neuroimaging (MRI, SPECT, fMRI, DTI), and neurophysiology (ERP, TMS, tDCS). In order to determine the functional outcome of cognitive disturbances following cerebellar damage and the underlying pathophysiological mechanisms, longitudinal follow-up studies are needed to disentangle the mysteries of this impressively competent structure at the bottom of the brain.

Acknowledgments The author would like to thank Kim van Dun, Hyo-Jung De Smet, Philippe Paquier, and Jo Verhoeven who contributed to parts of this chapter.

References

Aarsen, F. K., Van Dongen, H., Paquier, P., Van Mourik, M., & Catsman-Berrevoets, C. E. (2004). Long-term sequelae in children after cerebellar astrocytoma surgery. *Neurology, 62*, 1311–1316.

Alexander, M. P., Gillingham, S., Schweizer, T., & Stuss, D. T. (2012). Cognitive impairments due to focal cerebellar injuries in adults. *Cortex, 48*, 980–990.

Arasanz, C. P., Staines, W. R., Roy, E. A., & Schweizer, T. A. (2012). The cerebellum and its role in word generation: A cTBS study. *Cortex, 48*, 718–724.

Ardila, A., & Surloff, C. (2006). Dysexecutive agraphia: A major executive dysfunction sign. *International Journal of Neuroscience, 116*, 653–663.

Babinski, J. (1902). *Sur le role du cervelet dans les actes volitionnels necessitant une succession rapide de mouvements (1)(Diadocoeinesie)*. Paris: Masson.

Baillieux, H., De Smet, H. J., Dobbeleir, A., Paquier, P. F., De Deyn, P. P., & Mariën, P. (2010). Cognitive and affective disturbances following focal cerebellar damage in adults: A neuropsychological and SPECT study. *Cortex, 46*, 869–879.

Baillieux, H., De Smet, H. J., Lesage, G., Paquier, P., De Deyn, P. P., & Mariën, P. (2006). Neurobehavioral alterations in an adolescent following posterior fossa tumor resection. *The Cerebellum, 5*, 289–295.

Baillieux, H., Vandervliet, E. J., Manto, M., Parizel, P. M., De Deyn, P. P., & Mariën, P. (2009). Developmental dyslexia and widespread activation across the cerebellar hemispheres. *Brain and Language, 108*, 122–132.

Baron, J., Bousser, M., Comar, D., Soussaline, F., & Castaigne, P. (1981). Noninvasive tomographic study of cerebral blood flow and oxygen metabolism in vivo. *European Neurology, 20*, 273–284.

Beaton, A., & Mariën, P. (2010). Language, cognition and the cerebellum: Grappling with an enigma. *Cortex, 46*, 811–820.

Beeson, P., Rapcsak, S., Plante, E., Chargualaf, J., Chung, A., Johnson, S., et al. (2003). The neural substrates of writing: A functional magnetic resonance imaging study. *Aphasiology, 17*, 647–665.

Bellebaum, C., & Daum, I. (2011). Mechanisms of cerebellar involvement in associative learning. *Cortex, 47*, 128–136.

Blancart, R. G., Escrig, M. G., & Gimeno, A. N. (2011). Aphasia secondary to left cerebellar infarction. *Neurología (English Edition), 26*, 56–58.

Botez-Marquard, T., Léveillé, J., & Botez, M. (1994). Neuropsychological functioning in unilateral cerebellar damage. *Canadian Journal of Neurological Sciences, 21*, 353–357.

Brown, W., Eliez, S., Menon, V., Rumsey, J., White, C., & Reiss, A. (2001). Preliminary evidence of widespread morphological variations of the brain in dyslexia. *Neurology, 56*, 781–783.

Catsman-Berrevoets, C. E., & Aarsen, F. K. (2010). The spectrum of neurobehavioural deficits in the Posterior Fossa Syndrome in children after cerebellar tumour surgery. *Cortex, 46*, 933–946.

Chheda, M., Sherman, J., & Schmahmann, J. (2002). Neurologic, psychiatric, and cognitive manifestations in cerebellar agenesis. In *Neurology* (Vol. 7, p. A356). Philadelphia: Lippincott Williams & Wilkins.

Cook, M., Murdoch, B., Cahill, L., & Whelan, B. M. (2004). Higher-level language deficits resulting from left primary cerebellar lesions. *Aphasiology, 18*, 771–784.

Copland, D. A., Chenery, H. J., & Murdoch, B. E. (2000). Persistent deficits in complex language function following dominant nonthalamic subcortical lesions. *Journal of Medical Speech-Language Pathology, 14*, 379–390.

De Smet, H. J., Catsman-Berrevoets, C., Aarsen, F., Verhoeven, J., Marien, P., & Paquier, P. F. (2012). Auditoryperceptual speech analysis in children with cerebellar tumours: A long-term follow-up study. *European Journal of Paediatric Neurology, 16*, 434–442.

De Smet, H. J., & Mariën, P. (2012). Posterior fossa syndrome in an adult patient following surgical evacuation of an intracerebellar haematoma. *The Cerebellum, 11*, 587–592.

De Smet, H. J., Paquier, P. F., Verhoeven, J., & Mariën, P. (2013). The cerebellum: its role in language and related cognitive and affective functions, Brain and Language, 127, 334-342.

Desmond, J. E. (2001). Cerebellar involvement in cognitive function: Evidence from neuroimaging. *International Review of Psychiatry, 13*, 283–294.

Desmond, J. E., Gabrieli, J. D., Wagner, A. D., Ginier, B. L., & Glover, G. H. (1997). Lobular patterns of cerebellar activation in verbal working-memory and finger-tapping tasks as revealed by functional MRI. *Journal of Neuroscience, 17*, 9675–9685.

Doyon, J., Penhune, V., & Ungerleider, L. G. (2003). Distinct contribution of the cortico-striatal and cortico-cerebellar systems to motor skill learning. *Neuropsychologia, 41*, 252–262.

Eckert, M. A., Leonard, C. M., Richards, T. L., Aylward, E. H., Thomson, J., & Berninger, V. W. (2003). Anatomical correlates of dyslexia: Frontal and cerebellar findings. *Brain, 126*, 482–494.

Exner, C., Weniger, G., & Irle, E. (2004). Cerebellar lesions in the PICA but not SCA territory impair cognition. *Neurology, 63*, 2132–2135.

Fabbro, F., Moretti, R., & Bava, A. (2000). Language impairments in patients with cerebellar lesions. *Journal of Neurolinguistics, 13*, 173–188.

Fabbro, F., Tavano, A., Corti, S., Bresolin, N., De Fabritiis, P., & Borgatti, R. (2004). Long-term neuropsychological deficits after cerebellar infarctions in two young adult twins. *Neuropsychologia, 42*, 536–545.

Fawcett, A. J., & Nicolson, R. I. (1999). Performance of dyslexic children on cerebellar and cognitive tests. *Journal of Motor Behavior, 31*, 68–78.

Fiez, J. A., Petersen, S. E., Cheney, M. K., & Raichle, M. E. (1992). Impaired non-motor learning and error detection associated with cerebellar damage. *Brain, 115*, 155–178.

Finch, A. J., Nicolson, R. I., & Fawcett, A. J. (2002). Evidence for a neuroanatomical difference within the olivo-cerebellar pathway of adults with dyslexia. *Cortex, 38*, 529–539.

Gasparini, M., Piero, V. D., Ciccarelli, O., Cacioppo, M. M., Pantano, P., & Lenzi, G. L. (1999). Linguistic impairment after right cerebellar stroke: A case report. *European Journal of Neurology, 6*, 353–356.

Gottwald, B., Wilde, B., Mihajlovic, Z., & Mehdorn, H. (2004). Evidence for distinct cognitive deficits after focal cerebellar lesions. *Journal of Neurology, Neurosurgery & Psychiatry, 75*, 1524–1531.

Gourovitch, M. L., Kirkby, B. S., Goldberg, T. E., Weinberger, D. R., Gold, J. M., Esposito, G., et al. (2000). A comparison of rCBF patterns during letter and semantic fluency. *Neuropsychology, 14*, 353.

Grabowski, T. J., Frank, R., Brown, C., Damasio, H., Ponto, L., Watkins, G., et al. (1996). Reliability of PET activation across statistical methods, subject groups, and sample sizes. *Human Brain Mapping, 4*, 23–46.

Hassid, E. I. (1995). A case of language dysfunction associated with cerebellar infarction. *Journal of Neurologic Rehabilitation, 9*, 157–160.

Heilman, K. M., Coyle, J. M., Gonyea, F., & Geschwind, N. (1973). Apraxia and agraphia in a left-hander. *Brain: A Journal of Neurology, 96*, 21–28.

Hetherington, R., Dennis, M., & Spiegler, B. (2000). Perception and estimation of time in long-term survi`s of childhood posterior fossa tumors. *Journal of the International Neuropsychological Society, 6*, 682–692.

Holmes, G. (1922). Clinical symptoms of cerebellar disease and their interpretation. The Croonian lectures I. *Lancet, 1*, 177–1182.

Hubrich-Ungureanu, P., Kaemmerer, N., Henn, F. A., & Braus, D. F. (2002). Lateralized organization of the cerebellum in a silent verbal fluency task: A functional magnetic resonance imaging study in healthy volunteers. *Neuroscience Letters, 319*, 91–94.

Ito, M. (2008). Control of mental activities by internal models in the cerebellum. *Nature Reviews Neuroscience, 9*, 304–313.

Ivry, R. B., & Diener, H. (1991). Impaired velocity perception in patients with lesions of the cerebellum. *Journal of Cognitive Neuroscience, 3*, 355–366.

Ivry, R. B., & Keele, S. W. (1989). Timing functions of the cerebellum. *Journal of Cognitive Neuroscience, 1*, 136–152.

Ivry, R. B., & Richardson, T. C. (2002). Temporal control and coordination: The multiple timer model. *Brain and Cognition, 48*, 117–132.

Katanoda, K., Yoshikawa, K., & Sugishita, M. (2001). A functional MRI study on the neural substrates for writing. *Human Brain Mapping, 13*, 34–42.

Kempler, D., VanLancker, D., Marchman, V., & Bates, E. (1999). Idiom comprehension in children and adults with unilateral brain damage. *Developmental Neuropsychology, 15*, 327–349.

Leggio, M. G., Chiricozzi, F. R., Clausi, S., Tedesco, A. M., & Molinari, M. (2011). The neuropsychological profile of cerebellar damage: The sequencing hypothesis. *Cortex, 47(1)*, 137–144.

Leggio, M., Neri, P., Graziano, A., Mandolesi, L., Molinari, M., & Petrosini, L. (1999). Cerebellar contribution to spatial event processing: Characterization of procedural learning. *Experimental Brain Research, 127*, 1–11.

Leggio, M. G., Silveri, M. C., Petrosini, L., & Molinari, M. (2000). Phonological grouping is specifically affected in cerebellar patients: A verbal fluency study. *Journal of Neurology, Neurosurgery & Psychiatry, 69*, 102–106.
Leggio, M., Solida, A., Silveri, M., Gainotti, G., & Molinari, M. (1995). Verbal fluency impairments in patients with cerebellar lesions. In *Society for neuroscience abstracts* (Vol. 364.6).
Leggio, M., Tedesco, A., Chiricozzi, F., Clausi, S., Orsini, A., & Molinari, M. (2008). Cognitive sequencing impairment in patients with focal or atrophic cerebellar damage. *Brain, 131*, 1332–1343.
Leiner, H. C., Leiner, A. L., & Dow, R. S. (1986). Does the cerebellum contribute to mental skills? *Behavioral Neuroscience, 100*, 443–454.
Longcamp, M., Anton, J.-L., Roth, M., & Velay, J.-L. (2003). Visual presentation of single letters activates a premotor area involved in writing. *Neuroimage, 19*, 1492–1500.
Luciani, L. (1891). *Il cervelletto: Nuovi studi di fisiologia normale e patologica. Coi tipi dei successori Le Monnier*. Firenze: Le Monnier.
Manni, E., & Petrosini, L. (2004). A century of cerebellar somatotopy: A debated representation. *Nature Reviews Neuroscience, 5*, 241–249.
Manto, M., & Mariën, P. (2015). Schmahmann's syndrome - identification of the third cornerstone of clinical ataxiology. *Cerebellum & Ataxias, 2*(1). http://doi.org/10.1186/s40673-015-0023-1
Mariën, P., Baillieux, H., De Smet, H. J., Engelborghs, S., Wilssens, I., Paquier, P., et al. (2009). Cognitive, linguistic and affective disturbances following a right superior cerebellar artery infarction: A case study. *Cortex, 45*, 527–536.
Mariën, P., Brouns, R., Engelborghs, S., Wackenier, P., Verhoeven, J., Ceulemans, B., et al. (2008). Cerebellar cognitive affective syndrome without global mental retardation in two relatives with Gillespie syndrome. *Cortex, 44*, 54–67.
Mariën, P., de Smet, E., de Smet, H. J., Wackenier, P., Dobbeleir, A., & Verhoeven, J. (2013a). "Apraxic dysgraphia" in a 15-year-old left-handed patient: Disruption of the cerebello-cerebral network involved in the planning and execution of graphomotor movements. *The Cerebellum, 12*, 131–139.
Mariën, P., De Smet, H. J., Wijgerde, E., Verhoeven, J., Crols, R., & De Deyn, P. P. (2013b). Posterior fossa syndrome in adults: A new case and comprehensive survey of the literature. *Cortex, 49*, 284–300.
Mariën, P., Engelborghs, S., Fabbro, F., & De Deyn, P. P. (2001). The lateralized linguistic cerebellum: A review and a new hypothesis. *Brain and Language, 79*, 580–600.
Mariën, P., Engelborghs, S., Pickut, B. A., & De Deyn, P. P. (2000). Aphasia following cerebellar damage: Fact or fallacy? *Journal of Neurolinguistics, 13*, 145–171.
Mariën, P., Michiels, E., & De Deyn, P. P. (2003). Cognitive and linguistic disturbances in the posterior fossa syndrome in children: A diaschisis phenomenon? *Brain and Language, 87*, 162.
Mariën, P., Saerens, J., Nanhoe, R., Moens, E., Nagels, G., Pickut, B. A., et al. (1996). Cerebellar induced aphasia: Case report of cerebellar induced prefrontal aphasic language phenomena supported by SPECT findings. *Journal of the Neurological Sciences, 144*, 34–43.
Mariën, P., Verhoeven, J., Brouns, R., De Witte, L., Dobbeleir, A., & De Deyn, P. (2007). Apraxic agraphia following a right cerebellar hemorrhage. *Neurology, 69*, 926–929.
Martin, A., Haxby, J. V., Lalonde, F. M., Wiggs, C. L., & Ungerleider, L. G. (1995). Discrete cortical regions associated with knowledge of color and knowledge of action. *Science, 270*, 102.
Marvel, C. L., & Desmond, J. E. (2010). The contributions of cerebro-cerebellar circuitry to executive verbal working memory. *Cortex, 46*, 880–895.
Middleton, F. A., & Strick, P. L. (1997). Cerebellar output channels. *International Review of Neurobiology, 41*, 61–82.
Miller, N., Reddick, W. E., Kocak, M., Glass, J. O., Löbel, U., Morris, B., et al. (2010). Cerebellocerebral diaschisis is the likely mechanism of postsurgical posterior fossa syndrome in pediatric patients with midline cerebellar tumors. *American Journal of Neuroradiology, 31*, 288–294.

Molinari, M., Chiricozzi, F. R., Clausi, S., Tedesco, A. M., De Lisa, M., & Leggio, M. G. (2008). Cerebellum and detection of sequences, from perception to cognition. *The Cerebellum, 7*, 611–615.

Molinari, M., Petrosini, L., Misciagna, S., & Leggio, M. (2004). Visuospatial abilities in cerebellar disorders. *Journal of Neurology, Neurosurgery & Psychiatry, 75*, 235–240.

Moretti, R., Bava, A., Torre, P., Antonello, R. M., & Cazzato, G. (2002a). Reading errors in patients with cerebellar vermis lesions. *Journal of Neurology, 249*, 461–468.

Moretti, R., Torre, P., Antonello, R. M., Carraro, N., Zambito-Marsala, S., Ukmar, M. J., et al. (2002b). Peculiar aspects of reading and writing performances in patients with olivopontocerebellar atrophy. *Perceptual and Motor Skills, 94*, 677–694.

Murdoch, B. E. (2010). The cerebellum and language: Historical perspective and review. *Cortex, 46*, 858–868.

Murdoch, B. E., & Whelan, B.-M. (2007). Language disorders subsequent to left cerebellar lesions: A case for bilateral cerebellar involvement in language? *Folia Phoniatrica et Logopaedica, 59*, 184–189.

Neau, J. P., Anllo, E., Bonnaud, V., Ingrand, P., & Gil, R. (2000). Neuropsychological disturbances in cerebellar infarcts. *Acta Neurologica Scandinavica, 102*, 363–370.

Nicolson, R. I., & Fawcett, A. J. (2011). Dyslexia, dysgraphia, procedural learning and the cerebellum. *Cortex, 47*, 117–127.

Nicolson, R. I., Fawcett, A. J., Berry, E. L., Jenkins, I. H., Dean, P., & Brooks, D. J. (1999). Association of abnormal cerebellar activation with motor learning difficulties in dyslexic adults. *The Lancet, 353*, 1662–1667.

Nicolson, R. I., Fawcett, A. J., & Dean, P. (1995). Time estimation deficits in developmental dyslexia: Evidence of cerebellar involvement. *Proceedings of the Royal Society of London B: Biological Sciences, 259*, 43–47.

Nicolson, R. I., Fawcett, A. J., & Dean, P. (2001). Developmental dyslexia: The cerebellar deficit hypothesis. *Trends in Neurosciences, 24*, 508–511.

Papathanassiou, D., Etard, O., Mellet, E., Zago, L., Mazoyer, B., & Tzourio-Mazoyer, N. (2000). A common language network for comprehension and production: A contribution to the definition of language epicenters with PET. *Neuroimage, 11*, 347–357.

Pernet, C., Andersson, J., Paulesu, E., & Demonet, J. F. (2009). When all hypotheses are right: A multifocal account of dyslexia. *Human Brain Mapping, 30*, 2278–2292.

Petersen, S. E., Fox, P. T., Posner, M. I., Mintun, M., & Raichle, M. E. (1988). Positron emission tomographic studies of the cortical anatomy of single-word processing. *Nature, 331*, 585–589.

Petersen, S. E., Fox, P. T., Posner, M. I., Mintun, M., & Raichle, M. E. (1989). Positron emission tomographic studies of the processing of singe words. *Journal of Cognitive Neuroscience, 1*, 153–170.

Pollack, I. F. (1997). Posterior fossa syndrome. *International Review of Neurobiology, 41*, 411–432.

Rae, C., Harasty, J. A., Dzendrowskyj, T. E., Talcott, J. B., Simpson, J. M., Blamire, A. M., et al. (2002). Cerebellar morphology in developmental dyslexia. *Neuropsychologia, 40*, 1285–1292.

Raichle, M. E., Fiez, J. A., Videen, T. O., MacLeod, A.-M. K., Pardo, J. V., Fox, P. T., et al. (1994). Practice-related changes in human brain functional anatomy during nonmotor learning. *Cerebral Cortex, 4*, 8–26.

Rapcsak, S. Z., & Beeson, P. M. (2000). Agraphia. In B. Crosson, L. J. G. Rothi, & S. Nadeau (Eds.), *Aphasia and language: Theory and practice* (pp. 184–220). New York: Guilford.

Richter, S., Matthies, K., Ohde, T., Dimitrova, A., Gizewski, E., Beck, A., et al. (2004). Stimulus-response versus stimulus-stimulus-response learning in cerebellar patients. *Experimental Brain Research, 158*, 438–449.

Salman, M. S. (2002). Topical review: The cerebellum: It's about time! But timing is not everything-new insights into the role of the cerebellum in timing motor and cognitive tasks. *Journal of Child Neurology, 17*, 1–9.

Schlösser, R., Hutchinson, M., Joseffer, S., Rusinek, H., Saarimaki, A., Stevenson, J., et al. (1998). Functional magnetic resonance imaging of human brain activity in a verbal fluency task. *Journal of Neurology, Neurosurgery & Psychiatry, 64*, 492–498.

Schmahmann, J. D. (2004). Disorders of the cerebellum: Ataxia, dysmetria of thought, and the cerebellar cognitive affective syndrome. *The Journal of Neuropsychiatry and Clinical Neurosciences, 16*, 367–378.

Schmahmann, J. D. (2010). The role of the cerebellum in cognition and emotion: Personal reflections since 1982 on the dysmetria of thought hypothesis, and its historical evolution from theory to therapy. *Neuropsychology Review, 20*, 236–260.

Schmahmann, J. D., & Pandya, D. N. (1997). The cerebrocerebellar system. *International Review of Neurobiology, 41*, 31–60.

Schmahmann, J. D., & Sherman, J. C. (1998). The cerebellar cognitive affective syndrome. *Brain, 121*, 561–579.

Schweizer, T. A., Alexander, M. P., Susan Gillingham, B., Cusimano, M., & Stuss, D. T. (2010). Lateralized cerebellar contributions to word generation: A phonemic and semantic fluency study. *Behavioural Neurology, 23*, 31–37.

Schweizer, T. A., Levine, B., Rewilak, D., O'Connor, C., Turner, G., Alexander, M. P., et al. (2008). Rehabilitation of executive functioning after focal damage to the cerebellum. *Neurorehabilitation and Neural Repair, 22*, 72–77.

Silveri, M. C., Leggio, M. G., & Molinari, M. (1994). The cerebellum contributes to linguistic production A case of agrammatic speech following a right cerebellar lesion. *Neurology, 44*, 2047–2047.

Silveri, M. C., Misciagna, S., Leggio, M. G., & Molinari, M. (1997). Spatial dysgraphia and cerebellar lesion A case report. *Neurology, 48*, 1529–1532.

Silveri, M. C., Misciagna, S., Leggio, M. G., & Molinari, M. (1999). Cerebellar spatial dysgraphia: Further evidence. *Journal of Neurology, 246*, 312–313.

Spencer, R. M., Verstynen, T., Brett, M., & Ivry, R. (2007). Cerebellar activation during discrete and not continuous timed movements: An fMRI study. *Neuroimage, 36*, 378–387.

Steinlin, M., Imfeld, S., Zulauf, P., Boltshauser, E., Lovblad, K. O., Ridolfi, L. A., et al. (2003). Neuropsychological long-term sequelae after posterior fossa tumour resection during childhood. *Brain, 126*, 1998–2008.

Stoodley, C. J., & Schmahmann, J. D. (2010). Evidence for topographic organization in the cerebellum of motor control versus cognitive and affective processing. *Cortex, 46*, 831–844.

Stoodley, C. J., Valera, E. M., & Schmahmann, J. D. (2012). Functional topography of the cerebellum for motor and cognitive tasks: An fMRI study. *Neuroimage, 59*, 1560–1570.

Strelnikov, K. N., Vorobyev, V. A., Chernigovskaya, T. V., & Medvedev, S. V. (2006). Prosodic clues to syntactic processing – A PET and ERP study. *NeuroImage, 29*, 1127–1134.

Tavano, A., & Borgatti, R. (2010). Evidence for a link among cognition, language and emotion in cerebellar malformations. *Cortex, 46*, 907–918.

Tavano, A., Grasso, R., Gagliardi, C., Triulzi, F., Bresolin, N., Fabbro, F., et al. (2007). Disorders of cognitive and affective development in cerebellar malformations. *Brain, 130*, 2646–2660.

Tedesco, A. M., Chiricozzi, F. R., Clausi, S., Lupo, M., Molinari, M., & Leggio, M. G. (2011). The cerebellar cognitive profile. *Brain, 134*, 3672–3686.

Timmann, D., Drepper, J., Maschke, M., Kolb, F., Böring, D., Thilmann, A., et al. (2002). Motor deficits cannot explain impaired cognitive associative learning in cerebellar patients. *Neuropsychologia, 40*, 788–800.

Ullman, M. T. (2004). Contributions of memory circuits to language: The declarative/procedural model. *Cognition, 92*, 231–270.

Whelan, B.-M., & Murdoch, B. (2005). Unravelling subcortical linguistic substrates: Comparison of thalamic versus cerebellar cognitive-linguistic regulation mechanisms. *Aphasiology, 19*, 1097–1106.

Zettin, M., Cappa, S. F., D'amico, A., Rago, R., Perino, C., Perani, D., et al. (1997). Agrammatic speech production after a right cerebellar haemorrhage. *Neurocase, 3*, 375–380.

Chapter 10
A Word in the Hand: The Gestural Origins of Language

Michael C. Corballis

10.1 Introduction

The evolution of language, it has been suggested, may be "the hardest problem in science" (Christiansen and Kirby 2003, p. 1). In 1866, shortly after the publication of Darwin's (1859) *Origin of Species*, it was so controversial that the Linguistics Society of Paris banned all discussion of the origins of language, and shortly afterwards the Philological Society of London followed suit. This seemed to create something of a hiatus, and it was not really until around a century later that interest picked up again. Even so, the topic remains highly contentious to the point that one might wonder whether the ban should be reinstated. One widely held theory is that held by the dominant linguist of our time, Noam Chomsky, and supported by a number of archeologists and prehistorians. This theory might be termed the "big bang" theory, since it holds that language emerged in a single step and was restricted to our own species, *Homo sapiens*. In evolutionary terms this theory effectively denies any evolutionary precursors and is nonspecific as to the mechanisms of evolutionary change. In this chapter, I discuss this theory first, and suggest that it is deeply implausible. I then go on to consider an evolutionary account that is in many respects its opposite. In this alternative view language evolved from manual gestures, and its origins go far back in primate evolution.

M.C. Corballis (✉)
School of Psychology, University of Auckland, Auckland 1010, New Zealand
e-mail: m.corballis@auckland.ac.nz

M. Mody (ed.), *Neural Mechanisms of Language*, Innovations in Cognitive Neuroscience, DOI 10.1007/978-1-4939-7325-5_10

10.1.1 The "Big-Bang" Theory

Chomsky (2010) maintains that there were no languages prior to 100,000 years ago, and that when language did emerge it was the outcome of a sudden event, perhaps a mutation, in a single individual. This occurred in Africa sometime after 100,000 years ago and was disseminated through the population prior to the dispersal of humans from Africa some 60,000 years ago. In the most recent of Chomsky's various formulations, the essence of language and underlying thought is unbounded Merge, the process whereby elements are merged in recursive fashion to create what Chomsky has also called "discrete infinity"—the limitless number of possible structures that can be formed. Unbounded Merge is the basis of what has also been termed *universal grammar*, the common grammatical core of all human languages.

This is a simple solution because it requires only a single event, and obviates the need to seek precursors to language in other species, or even in our hominin ancestors. In part, its rationale follows from the very notion of unbounded Merge, which according to Chomsky must have arisen in all-or-none fashion. Just as the notion of infinity itself was not achieved by the ability to count to increasingly large numbers, so it seems highly unlikely that unbounded Merge could be reached by gradually extending the bounds on progressive Merge operations. As we shall see, though, this conclusion can be challenged.

Nevertheless the big-bang theory receives independent support from evidence that a "cultural revolution" took place in human prehistory within the past 100,000 years. It was characterized by the seemingly abrupt appearance of what has been termed "modern" behavior, including the fashioning of bodily ornamentation from shells, beads, or animal teeth, the emergence of sophisticated cave art, and a sudden advance in the level of technology. Summarizing the evidence, the archaeologist Sir Paul Mellars concludes as follows:

> To describe the Upper Paleolithic revolution in Europe as reflecting preeminently an explosion in explicitly symbolic behavior and expression is in no sense an exaggeration, as most prehistorians would now agree. We are probably on safe ground in assuming that symbolic behavior and expression of this level of complexity would be inconceivable in the absence of highly structured language systems and brains closely similar, if not identical to, our own (Mellars 2005, p. 12).

Although this extract seems to imply that the revolution took place in Europe, most now agree that its beginnings were evident in Africa prior to the dispersal, but it is still argued that it occurred well after the appearance of anatomically modern humans, dated at around 200,000 years ago (Marean 2010).

In a recent popular book Tattersall (2012) affirms that the cultural revolution did indeed coincide with the big bang that gave us language:

> Our ancestors made an almost unimaginable transition from a non-symbolic, nonlinguistic way of processing information and communicating information about the world to the symbolic and linguistic condition we enjoy today. It is a qualitative leap in cognitive state unparalleled in history. Indeed, as I've said, the only reason we have for believing that such a leap could ever have been made, is that it *was* made. And it seems to have been made well *after* the acquisition by our species of its distinctive modern form (p. 199).

In its apparent simplicity, though, the big-bang theory smacks of the miraculous. The tone of the above extract does imply that there is no real precedent, nor any clear explanation, for the "unimaginable transition." Its rather biblical quality suggests that it may have been driven as much by wishful thinking as by scientific evidence. It serves, for example, to justify the manner in which we exploit other animals, and perhaps exonerate us from the suspicion that our forebears may have been responsible for the demise of our close but now extinct relatives the Neanderthals and Denisovans, with whom we coexisted until some 30,000 years ago.

More importantly, the big-bang theory is profoundly at odds with Darwin's theory of evolution by natural selection. Darwin himself wrote:

> If it could be demonstrated that any complex organ existed, which could not possibly have been formed by numerous, successive, slight modifications, my theory would absolutely break down. But I can find no such case (Darwin 1859, p. 158).

If correct, then, the big-bang theory of language evolution might be the very case that Darwin feared. Chomsky himself has often referred to language as "an organ of the body" (e.g., Chomsky 2007, p. 5), at least comparable in complexity to, say, the heart or liver. So far as we know, no such organ has emerged as a result of a single mutation in a single individual.

10.1.2 The Case for Continuity

Pinker and Bloom (1990) wrote in support of a Darwinian account of language evolution in a landmark article and included the following comment:

> If a current theory of language is truly incompatible with the neo-Darwinian theory of evolution, one could hardly blame someone for concluding that it is not the theory of evolution that must be questioned, but the theory of language (p. 708).

As a first step, then, one can challenge the idea that language, by its very nature, must have evolved in all-or-none fashion. Unbounded Merge is really an idealized abstraction, since people do have limits on the extent to which they can merge linguistic elements, whether in production or comprehension. One example of the Merge operation is the embedding of clauses within sentences, as in *The rat that the cat killed ate the malt.* We can push it to another level of embedding, as in *The rat that the cat that the dog chased killed ate the malt*, although this borders on unintelligibility. Yet another level taxes the system to an unacceptable degree: *The rat that the cat that the dog that the cow worried chased killed ate the malt*. Contrary to the notion of unbounded Merge, Fujita (2009), among others, has proposed a gradual evolution of Merge through progressive levels of recursion. Moreover some languages, such as those of the Pirahã in Brazil (Everett 2005) and the Iatmul of New Guinea (Evans 2009), appear to have no recursive embedding at all. The sheer diversity among the world's 7000 or so languages has led several to claim that the very notion of universal grammar, as proposed by Chomsky, has no validity (e.g., Christiansen and Kirby 2003; Evans and Levinson 2009; Tomasello 2008).

The evidence for the cultural revolution underlying the big bang has also been challenged. Marean et al. (2007) report evidence for "modern" human activity going back 160,000 years, close to the date at which *H. sapiens* emerged and well before the supposed big bang. McBrearty and Brooks (2000) write of the "revolution that wasn't," suggesting a more gradual rise in technological sophistication from the Middle Stone Age around 250,000–300,000 years ago. More recently Shea (2010) has argued that the apparent spurt of technology supposedly evident in Africa in the Middle Paleolithic was more likely a phase that came and went, an indication of variability rather than a sudden and universal shift toward greater sophistication. In much more recent times, the colonizers of the Americas or the European explorers in Africa found little if anything to resemble the burst of symbolic behavior in the Upper Paleolithic, as described by Mellars (2005).

A corollary of the big-bang theory is that the Neanderthals and Denisovans would not have possessed the faculty of language. These large-brained hominins all shared a common ancestor with *H. sapiens* dating from some 500,000 years ago, and were separated from around 400,000 years ago, with the lineages leading to Neanderthals and Denisovans in Europe and that leading to modern humans confined to East Africa until the dispersal from around 60,000 years ago. The wandering humans eventually made contact with their long-lost cousins in Europe and Russia from around 40,000 years ago. Sequencing of both the Neanderthal (Green et al. 2010) and Denisovan (Meyer et al. 2012) genomes reveals a degree of interbreeding both with each other and with anatomically modern humans. This suggests in turn that the three species had common cognitive capacities, and indeed raises questions as to whether they were different species (Reich et al. 2010). At worst, the Neanderthals and Denisovans might have had some linguistic deficiencies relative to *H. sapiens,* but not the total lack implied by the big-bang theory.

Several arguments have been mounted to show that Neanderthals were incapable of language, but have proven far from definitive (see Johansson 2013 for review). Suggestions that the Neanderthal vocal tract could not sustain articulate speech (e.g., Lieberman 2007) have proven at best inconclusive (e.g., Barney et al. 2012; Boë et al. 2007). A mutation of the FOXP2 gene resulted in a severe speech impediment in about half the members of an extended English family, leading to suggestions that a (different) mutation of this otherwise highly conserved gene may have occurred in the human lineage to enable speech (Enard et al. 2002). This appears to have been refuted by the discovery that the region on the human version of the gene thought to be critical was present in Neanderthal DNA (Krause et al. 2007)—although the human FOXP2 may still differ in other ways (Ptak et al. 2009).

The question of whether or not the Neanderthals were capable of articulate speech in any case need not bear on the broader question of whether they were capable of language. Even in Chomskyan theory, the internal basis of language is symbolic and amodal; external language (E-language) interfaces with I-language, and can be expressed either vocally or manually, or a mixture of the two. Indeed, it may well be that external language was first expressed through gesture rather than speech, and the Neanderthals communicated in language-like ways with their hands rather than their voices—or more likely both. In the following sections I argue that language did evolve from manual gestures, although the transition to speech was probably gradual, and indeed incomplete.

10.2 The Gestural Theory

Speculation that language evolved from manual gestures has a long history. One early proponent was the philosopher and cleric de Condillac (1971). He was an ordained priest, and was afraid of offending the Church's view that language was a gift from God, so he disguised his theory as a fable about two children abandoned in the desert after the Flood. At first they communicated by using bodily movements, until eventually replacing their gestures with vocal sounds. Over the succeeding centuries many others have made similar suggestions as to the priority of manual gesture, including such dominant figures as Rousseau (1782), Wundt (1900), and the neurologist Critchley (1939, 1975).

A more comprehensive case for gestural origins was proposed by Hewes et al. (1973), and set the stage for subsequent developments (e.g., Arbib 2005; Corballis 1991, 2002; Rizzolatti and Sinigaglia 2008; Tomasello 2008). It presents a very different perspective on language evolution from the big-bang theory.

10.2.1 The Primate Mirror System

A convenient starting point, but one unknown to Hewes, was the discovery of mirror neurons in area F5 of the rhesus monkey (Rizzolatti et al. 1988). Area F5 is a region in the prefrontal cortex homologous to Broca's area in the human brain. Mirror neurons are so called because they are active not only when the animal executes a hand movement to grasp an object, but also when it observes another individual making the same movement. They therefore represent a mapping, or mirroring, between perception and action, thereby suggesting an evolutionarily early template for the later emergence of a communication system. Mirror neurons are now understood to be part of a more general mirror system, incorporating the superior temporal sulcus and area PF in the inferior parietal lobule (Rizzolatti and Sinigaglia 2010). Area PF is homologous to part of Wernicke's area in the human brain.

It is now known too that the homologous areas in humans mediate sign language, as well as symbolic gestures such as pantomimes and emblems (e.g., Xu et al. 2009). Mirror neurons tuned to manual grasping, whether executed or observed, have also been confirmed from single-cell recordings in humans with intractable epilepsy and were located in a variety of areas, including medial frontal and temporal regions (Mukamel et al. 2010).

Arbib (2005) has speculated in detail as to how language might have evolved from the mirror system. In primate evolution, he suggests, the mirror system progressed from a system for grasping to one that included imitation, and from there to a system supporting sequences of actions, enabling pantomime. This was then the substrate for what Arbib called "protosign," an open-ended system of signs, but without the combinatorial, syntactic structure of language. This in turn provided a scaffold for protospeech, which up to that stage owed little to nonhuman vocalizations. From that point, an "expanding spiral" of protosign and protospeech led to true language, whether spoken or signed. While this seems a plausible scenario, it is also possible that gestural language evolved to a syntactic level similar to that of modern signed languages before vocalization was introduced. Tomasello (2008), for example, writes of the possibility "that the human capacity evolved quite a long way in the service of gestural communication alone, and the vocal capacity is actually a very recent overlay (p. 246)."

In this view, then, the origins of language lie in a system initially dedicated to manual grasping, and indeed the concept of grasping underlies many expressions to do with language itself. The word *grasp* is often used to mean "understand." *Comprehend* and *apprehend* derive from Latin *prehendere,* "to grasp"; *intend, contend,* and *pretend* derive from Latin *tendere*, "to reach with the hand"; we may *press* a point, and *expression* and *impression* also suggest pressing. We *hold* conversations, *point* things out, *seize upon* ideas, *grope for* words. It works visually, too, as when you *see* what somebody means. Fonagy and Target (2007, p. 437) suggest that such examples are indeed "a residue of gestural language."

To contribute to communication, the mirror system in nonhuman primates required at least one further modification. Mirror neurons in primates are *transitive* in that they respond only to actions involving the grasping of an actual object, even if that object is hidden behind a screen but the animal knows it is there (Umilta et al. 2001). In humans, evidence from neuroimaging suggests that the mirror system responds as well to *intransitive* actions, where a movement occurs in the absence of an object to be grasped. This suggests that the human mirror system was extended to incorporated actions that were symbolic along with those that were object related. This may have come about, not through a "big bang" mutation providing for symbolic representation, but from the requirement of language to refer to events not present in the here-and-now (Corballis 2009). This is almost tangible in American Sign Language where, in conversation, absent objects are notionally "placed" at different locations in front of the speaker, and reference to those objects is achieved by pointing to the appropriate location (Emmorey 2001). Speakers often point in a similar way when they gesture as they talk (Kendon 2004).

In nonhuman primates, the mirror system does not respond to vocalization (Rizzolatti and Sinigaglia 2008), which probably reflects the involuntary nature of primate vocalization. This is a further reason to suppose that language evolved from manual actions, since language is an intentional system. Kohler et al. (2002) found mirror neurons in monkeys to be receptive to nonvocal acoustic input, including the sounds of manual acts such as tearing paper or cracking peanuts, but there was no response to monkey calls. In humans, perhaps uniquely, vocalization does appear to have been incorporated into the mirror system and at the same time become amenable to volition. This may have required a direct connection between the motor cortex and the nucleus ambiguus, the subcortical structure responsible for the coordinated movements involved in speech (Jürgens 2002).

The incorporation of vocalization into the mirror system is captured in the motor theory of speech perception, which holds that speech sounds are perceived in terms of how they are articulated rather than in terms of their acoustic properties (e.g., Galantucci et al. 2006). Nevertheless the motor theory does not completely explain speech perception. Hickok (2009) points out that people with Broca's aphasia lose the ability to produce articulate speech yet can still understand it, and human infants also understand words before they can produce them. However, in a recent paper revisiting the perception of the speech code, Fowler et al. (2016) clarify that phonetic segments are articulatory and coarticulation does not eliminate them; rather, phonetic segments appear to leave an acoustic signature that listeners can track.

10.2.2 Dual-Stream Theory

Hickok and Poeppel (2007) have suggested a "dual-stream" account of speech processing in humans, with a ventral stream for comprehension and a dorsal stream mapping onto the frontal-lobe articulatory networks. It is only the dorsal stream, then, that involves the mirror system, one role of which is to shape action to be compatible with input. In nonhuman primates, this role is well advanced for manual action, but not for vocal output. In humans, though, the mirror system appears to incorporate vocal production, and allow for productive language to emerge in the vocal modality. Both dorsal and ventral systems, moreover, appear to be involved in syntax, in different ways.

Van der Lely and Pinker (2014) draw attention to a subtype of specific language impairment (SLI) in which the primary deficit is grammatical. Children with this subtype, known as Grammatical-SLI, are especially deficient in processing complex grammatical constructions. To account for this deficiency, van der Lely and Pinker distinguish between basic and extended syntax. Basic syntax connects words to meanings, and deals with short-range syntax, as in words in which tense is conveyed in the word as a whole, such as the English irregular past-tense forms "ate" or "bought." Basic syntax is represented in the ventral system and may be sufficient for what has been termed protolanguage—language without combinatorial syntax. Extended syntax is combinatorial, as in the merging of morphemes to represent past tense ("jumped," "climbed") or in more complex cases where understanding may depend upon integrating across different words, as in sentences where subject and verb are separated by intervening clauses. Extended syntax depends on the dorsal system, left-hemispheric in most people, and involves interaction between lateral frontal cortex and basal ganglia (Mestres-Missé et al. 2012).

A possibly related dissociation may occur in sign language. Hickok et al. (2009) showed that when narrating a story in American Sign Language, signers with left-hemisphere damage make more lexical than classifier errors, whereas right-hemisphere damage results in the reverse pattern. Lexical signs signal relations in categorical form, where classifier signs are more analogue in character, and of debatable linguistic status. The authors suggest that the production of lexical signs is strongly left-hemispheric, whereas producing classifier forms relies on both hemispheres. In this study the brain damage was located primarily in fronto-parietal regions, and in two cases the left-hemisphere damage, resulting primarily in lexical errors, was in the basal ganglia. Again, then, it may be the left-hemispheric dorsal system, which includes the basal ganglia, that underlies the grammatical component, and that emerged from the primate mirror system.

10.2.3 The Role of Learning

Language is not only intentional, it also requires learning. Each of the some 7000 different languages in the world requires a distinctive set of outputs to be learned. Of course most of these languages are spoken, requiring the learning of vocal patterns. Although vocal communication is widespread in the animal kingdom, surprisingly few species are capable of vocal learning. Nonhuman primates are conspicuously absent from the list, which includes elephants, seals, killer whales, and some birds (Jarvis 2006). Jürgens (2002) links failure of vocal learning to the absence of cortical representation:

> As the vocalizations of monkeys, cats, and bats are almost completely genetically determined in their acoustic structure, while the vast majority of human vocalizations are more or less completely learned, the difference in cortical representation might reflect the different role motor learning plays in vocal behavior of these species (p. 246).

Jarvis and Jürgens concur that among the existing primates, only humans are vocal learners.

Nevertheless some studies suggest a degree of modifiability in some primate calls, although this may have to do with subtle changes within call types rather than the generation of new call types (Egnor and Hauser 2004). Chimpanzees modify their screams when under attack, depending on the severity of the attack and their status relative to that of nearby chimps (Slocombe et al. 2010). When encountering food, chimps also emit different kinds of grunts depending on the type of food (Slocombe and Zuberbühler 2005). Some modifications involve the face and mouth rather than voicing itself. For instance chimpanzees can modify vocal sounds to attract attention by vibrating their lips, as in the "raspberry" sound (Hopkins et al. 2007), and this call can be imitated by naïve animals in captivity (Marshall et al. 1999).

In a recent review Petkov and Jarvis (2012) suggest that the distinction between vocal learners and vocal nonlearners may not be absolute, but may be a matter of degree. The extraordinary capacity of humans to learn complex and vastly different speech patterns is nevertheless well beyond that of any other known primate. It would not be surprising, though, if precursors of vocal learning did exist in our more recent primate forebears, since evolutionary change does not emerge out of nothing. Even so, there seems little doubt that the capacity to learn manual patterns of activity goes far back in primate evolution.

10.2.4 Communication in Great Apes

The clearest evidence that intentional manual action moves beyond grasping to serve an intentional communicative function in nonhuman primates comes from the great apes. No attempt to teach apes to speak has proven even remotely successful (e.g., Hayes 1951; Kellogg and Kellogg 1933; Ladygina-Kohts et al. 2002), but moderate success has been achieved using forms of sign language. One of the first to achieve some success was the chimpanzee Washoe, using a simplified form of American Sign Language (Gardner and Gardner 1969). More impressive is the bonobo Kanzi, who communicates by pointing to arbitrary signs representing objects and actions to make simple requests (Savage-Rumbaugh et al. 1998). The keyboard Kanzi uses has over 300 signs, and he supplements these by inventing gestures of his own. The gorilla Koko is said to use and understand over 1000 signs (Patterson and Gordon 2001).

These examples are somewhat artificial, since they are drawn from animals in artificial settings created by humans, but observations of apes in the wild also suggest a dominance of bodily over vocal communication, especially where communication is intentional rather than emotional or instinctive. Pollick and de Waal (2007) compared manual gestures directly with orofacial movements and vocalizations in the natural communications of captive chimpanzees and bonobos, and found manual gestures to be much less tied to context, and more variable between groups. The relative flexibility of manual gestures was more pronounced in bonobos than in chimpanzees, and only in bonobos did the combining of facial and vocal signals to manual gestures add to the impact on the recipient. This study is not definitive with respect to intentional control over vocalization, since vocalizations were lumped together with orofacial movements, and many such movements in chimpanzees and bonobos, such as lipsmacks, are not vocalized, but may well be under intentional control. This study nevertheless confirms the dominance of manual gesture in the natural communications of our closest nonhuman relatives.

In perhaps the most comprehensive study of chimpanzee gestures in the wild, Hobaiter and Byrne (2011) spent 266 days of recording gestures made by chimpanzees in the Budongo National Park in Uganda. Gestures included movements of the body, limbs, and head, but excluded facial expressions. They were intentional in that they were directed to another chimp, with the apparent aim of influencing the receiver's behavior, and they were also described as "mechanically ineffective," presumably to exclude acts like fighting, eating, or manipulation of objects. They included actions like a directed push, a handshake, an embrace, sexual display, and what has been called the "directed scratch," in which the chimp scratches the part of the body where he or she wants to be groomed by another chimp (Pika and Mitani 2009). It has been suggested that grooming itself is precursor to language (Dunbar 1998).

A total of 4397 gestures were recorded, made up of at least 66 different kinds of gestures. Most were recorded during play, but others occurred during other activities, such as grooming, travelling, or feeding. Just over 60% of them were single gestures, but others occurred in bouts, most commonly made up of two gestures, but in sequences of three, four, or at most five, with decreasing frequency. These bouts did not seem to resemble sentences in any way. Instead they seemed to be largely redundant, as though trying to get a single point across in different ways.

It has also been proposed that tool making is a precursor to language (Stout and Chaminade 2012), and there are well documented examples of chimpanzees making and using tools. They fashion sticks for fishing termites out of holes (Bogart and Pruetz 2008) and make spears for jabbing into the hollow trunks of trees to extract bush-babies and then eat them (Pruetz and Bertolani 2007). Chimpanzees in the Laongo National Park in Gabon use tool sets comprising up to five different stick and bark tools to extract honey from hives (Boesch et al. 2009). Some 25 different kinds of chimpanzee tools have been documented.

Of course chimpanzees do vocalize, but their vocalizations are largely involuntary, perhaps even exclusively so. Premack (2007), a pioneer in the study of chimpanzee behavior, notes that "chimpanzees lack voluntary control of their voice" (p. 13,866). Another pioneering observer of chimpanzees once wrote that "(t)he production of sound in the absence of the appropriate emotional state seems to be an almost impossible task for a chimpanzee" (Goodall 1986, p.125). By the same token, emotionally induced primate vocalization may be difficult to suppress. Goodall writes of a chimpanzee trying to stifle a call induced by the discovery of food by placing a hand over his mouth—evidence of the intentional dominance of a manual gesture over an involuntary vocal call. The capacity for intentional control of the voice, largely absent in the chimpanzee, must therefore have evolved after the separation of the hominin line from that leading to modern chimpanzees.

10.2.5 Understanding of Speech, Without Production

Even great apes, then, seem incapable of producing speech-like sounds, but some have shown remarkable ability to understand speech. Savage-Rumbaugh et al. (1998) reported that Kanzi was able to follow instructions, made up of several spoken words, at a level comparable to that of two-and-a-half-year-old child. Kanzi is now said to understand some 3000 words (Raffaele 2006). The gorilla Koko, too, can respond meaningfully to simple spoken requests, but cannot produce anything resembling speech (Patterson and Gordon 2001). The ability to recognize spoken words, and even to understand combinations, is not restricted to apes. A border collie known as Rico responds accurately to spoken requests to fetch different objects from another room and then either to place the designated object in a box or to bring it to a particular person (Kaminski et al. 2004). In what is shaping as a linguistic dog fight, another border collie called Chaser is said to know the spoken names of 1022 objects (Pilley and Reid 2011).

These observations support Hickok and Poeppel's suggestion of a ventral system of vocal understanding not involving the mirror system. What chimpanzees and dogs lack is not the ability to understand symbols, but rather the ability to produce vocalizations to act as symbols, as well as the capacity for extended syntax. Their ability to understand spoken words may derive from the need to respond to complex sounds, whether emanating from the calls of other animals or from other environmental events, such as thunder or the sound of a tree falling—or the sound of another species. In short, the auditory system may well have been pre-adapted for the receptive aspects of language well before our forebears were able to produce language-like sounds. It is with respect to the production of language, then, that the vocal system in nonhuman species falls short, suggesting that language evolved initially within the visuo-manual mode, which was already preadapted for both the perception and production of intentional acts.

The ability of apes and dogs to respond meaningfully to spoken commands also suggests that symbolic representations are not restricted to humans, as implied by the big-bang theory, and indeed by those authors cited earlier who suggest that a cultural revolution occurred because of a new-found symbolic understanding. The burden of understanding the evolution of human language falls not on how symbols were invented, but rather on how output systems evolved to create intentional signals—and perhaps also on how these systems evolved the combinatorial capacity to create complex meaning.

10.2.6 Human Gestures

Part of the argument that language evolved from gesture is that the hands play an important but often neglected role in the way people communicate. As long ago as the first century AD, the Roman rhetorician Quintilian wrote:

> As for the hands, without which all action would be crippled and enfeebled, it is scarcely possible to describe the variety of their motions, since they are almost as expressive as words (Quintilian 1920, pp. 290–291).

It has also long been known that signed languages can serve in place of speech, especially among the deaf and mute, as well as in some speaker-hearer communities, notably in aboriginal Australia (Kendon 2004) and among some native North American groups (Mithun 1999).

It was not really until the mid-twentieth century, though, that sign language was understood to be the equal of spoken language in grammatical and semantic sophistication, and even the manual equivalents of phonology (Emmorey 2001), to the point that American Sign Language (ASL) was declared to be the official language of Gallaudet University, a university for the deaf in Washington DC. This was largely due to William C. Stokoe, who taught at Gallaudet and persuaded the university to switch from oral methods. Stokoe used his understanding and knowledge of sign language to argue that language evolved from manual gestures (Armstrong et al. 1995; Stokoe 2001).

The very nature of sign language provides clues as to how language might have emerged. The hands provide a natural medium for representing objects and actions in the world, with the movement of the hands and arms describing trajectories and the hands taking the shapes of objects. Emerging sign languages typically begin with pantomime, but signs are then conventionalized so that many no longer provide a pictorial indication of what they stand for (Burling 1999). Conventionalization may be at the cost of transparency, but leads to great efficiency. On an evolutionary scale, speech itself may be the end product of a conventionalization process that began with pantomime.

The distinction between speech and gestural communication is in any case not absolute, since speech is universally accompanied by manual gestures, and the tight synchrony between the two suggests that they are controlled by a single integrated system (McNeill 1985). Experiments show that gestures influence the understanding of speech, just as speech influences the understanding of gestures, an interaction that is mutual and obligatory. Describing these results, Kelly et al. (2009) remark that speech and gesture are "two sides of the same coin." Modern language ranges from pure speech, as on radio or telephone, to pure manual gesture, as in signed languages, with conversational language somewhere in between. Moreover, if prevented from speaking, people naturally invent gestural schemes, which can take on grammatical properties (Goldin-Meadow et al. 1996).

The co-occurrence of manual gesture and speech, and perhaps the evolutionary transition of dominance from one to the other, can be understood more readily if speech itself is regarded as a gestural system, comprising movements of the lips, the larynx, the velum, and the blade, body and root of the tongue (Studdert-Kennedy 2005). In the course of evolution, intentional communication may have evolved from manual gestures, to overt facial gestures, and finally to the largely hidden gestures that comprise speech—although all three forms of gesture remain present. Speech gestures, although largely contained within the mouth, retain a visible component, as illustrated by the McGurk effect: A syllable (such as *da*) is dubbed onto a mouth saying another syllable (such as *ba*), and people tend to "hear" what they see rather than what was actually voiced (McGurk and MacDonald 1976). Other studies show the parts of the brain involved in producing speech are activated when people simply watch silent videos of people speaking (Calvert and Campbell 2003; Watkins et al. 2003). Ventriloquists know the power of vision over what one hears when they project their own voices onto the face of a dummy by synchronizing the mouth movements of the dummy with their own tight-lipped utterances.

The retreat of gestures into the mouth may be regarded as an early example of *miniaturization*, freeing the hands and the rest of the body for other activities.

10.2.7 Handedness and Brain Asymmetry

Among the arguments mounted by Hewes et al. (1973) for the gestural origins of language was the coincidence of handedness and brain asymmetry:

> The peculiarly human association of right-handedness and left-hemisphere dominance for both language skill and precise manual manipulations could well be the outcome of a long selective pressure for the clear separation of the precision grip from the power grip, combined with a manual-gesture language exhibiting a similar (and related) asymmetry (p. 9).

Most people are indeed right-handed and left-cerebrally dominant for speech, although some individuals are left-handed or right-cerebrally dominant for speech. The correlation, though, is actually fairly weak (Badzakova-Trajkov et al. 2010), perhaps because handedness, in particular, is subject to random influences (Annett 2002; McManus 2002).

These asymmetries appear to be present in the chimpanzee, although they are less pronounced than in humans. Evidence suggests that a majority of chimpanzees are right-handed, both in captivity (Hopkins 1996; Hopkins and Leavens 1998) and in the wild (Biro et al. 2006; Boesch 1991; Humle and Matsuzawa 2009; Lonsdorf and Hopkins 2005), although the incidence is round 65–70%, lower than the 90% incidence in humans. Moreover, in the majority of chimpanzee brains, the temporal planum, considered partly homologous to Wernicke's area in humans, is larger on the left than on the right (Gannon et al. 1998; Hopkins et al. 1998)—and asymmetry apparently absent in rhesus monkeys and baboons (Wada et al. 1975) but well documented in humans (e.g., Foundas et al. 1996). This leftward asymmetry in the chimpanzee is correlated with a right-handed bias in gestural communication (Hopkins and Nir 2010).

Since the leftward bias is evident in great apes and is associated with manual actions and not with vocal output, it again supports a priority for manual gesture rather than vocal calls in the evolution of language. Evidence also reveals that, in humans, the cortical control of skilled manual action, or praxis, is biased to the left hemisphere in both left- and right-handers, suggesting a stronger relation to language asymmetry than handedness itself (Vingerhoets et al. 2012). This further supports the idea that language evolved within a system specialized for manual control of skilled movement that was already biased toward the left hemisphere (Corballis et al. 2012).

10.2.8 Priority of Gesture in the Development of Language

It has become increasingly apparent that gesture comes before speech in the development of language in human children. Of course, the adage that "ontogeny recapitulates phylogeny," attributed to Ernst Haeckel, has been largely discredited (Gould 1977), but the sequence of events in the development of language can nevertheless provide clues as to how language might have come about in evolution.

In infants, intentional communication with manual gestures is evident from as early as 10 months. Pointing is especially dominant early on and is used both to request things, and to express attention to things (Tomasello 2008). Spoken words are gradually introduced, but even at 16 months of age manual gestures are more frequent. By 20 months, the balance has shifted, and spoken words become slightly more frequent. The most frequent two-word utterances, though, nearly always combine a gesture and a word, and it is not until later that two-word utterances are composed of two spoken words (see Capirci and Volterra 2008 for review). Manual gestures in early childhood can also predict later language success even up to the two-word level (Iverson and Goldin-Meadow 2005).

Even in 1-year-olds, pointing seems to have a property that is unique to humans. Chimpanzees can point to objects that are out of reach, with the aim of having a person hand the object to them, but virtually all of these pointing gestures are imperative, aimed at gratification. In contrast, 1-year-old children can point with the apparent objective of sharing information, rather than of gaining reward. Tomasello (2008) gives a number of examples where the intention is to share rather than request gratification. A 13-month-old child watches as her father arranges the Christmas tree. Her grandfather then enters and the child points to the tree for him, as if to say "Look!" At 13.5 months, while her mother is looking for a missing refrigerator magnet, a child points to a basket of fruit, under which the magnet is hidden.

Gestures therefore precede speech in development and serve to scaffold the development of speech. One might conjecture that gesture played a similar role in the evolution of language. Indeed, the role of gesture may have been even more critical in language evolution, since the early hominins, unlike human infants, may have been more severely lacking in the capacity to produce intentional vocalization.

10.3 An Evolutionary Scenario

The seed for language may have been sown by the development of intentional manual action in our primate forebears. These animals lived in forested environments where the hands were extensively employed in climbing and swinging from branches, as well as in such activities as plucking fruits, catching insects, and grooming. The mirror system evolved to map the perception of these actions onto their production. Although it has also been suggested that the mirror system evolved primarily to mediate imitation (e.g., Rizzolatti and Craighero 2004), monkeys are poor imitators, and the system probably had more to do with shaping action to conform to perception. Even plucking a piece of fruit requires the act to be conducted in accordance with perception, and the mirror system may have emerged primarily to mediate the monkey's calibration of its own actions, rather than those of another.

Nevertheless the system does respond to actions of other individuals, perhaps more by extension than original design, laying the groundwork for intentional communication. In great apes, the intentional motor system seems developed to the point of tool use and manufacture, as well as gestural communication. In interactions with humans, at least in artificial settings, their gestures have some language-like properties, including a degree of combinatorial capacity, reference to nonpresent objects, and the construction and understanding of simple commands. It is widely agreed that no great ape has yet demonstrated true grammatical language, although this view is at least partly dependent on Chomskyan views on the nature of language. Even so, there is little evidence from chimpanzees or bonobos of a Merge operation, unbounded or not.

The emergence of bipedalism in hominins probably enhanced the complexity of gestures as well as of the use and manufacture of tools. The critical phase was probably the Pleistocene, dating from around 2.9 million to about 12,000 years ago. The Pleistocene saw the beginning of a tripling of brain size, the emergence of stone tools, and obligate replacing the more facultative bipedalism of the earlier hominins. The Pleistocene is also widely recognized as the period in which hominins came to occupy what has been termed the "cognitive niche" (Tooby and de Vore 1987), depending on social bonding and enhanced communication for survival in the more exposed and dangerous environment of the African savanna. It seems highly likely, then, that grammatical language evolved gradually during this era, rather than late and abruptly, as implied by the big-bang theory.

As Donald (1991) argued, language probably emerged from pantomime, using the body to mimic space-time activities and so convey them to a watching audience. Pantomime, though, is inefficient, and over the course of the Pleistocene, the pressure toward a more efficient and compact system may have driven the process of conventionalization. Iconic or pantomimic gestures were replaced by simpler signals whose meanings were acquired through association rather than through pictorial representation. Meaning is then carried through cultural transmission, rather than in the signal itself. Such signals might be described as arbitrary symbols, but their arbitrary nature arose from practical concerns rather than from some newfound cognitive capacity—recall that chimpanzees and border collies can learn to associate spoken words with objects and actions. As I suggested earlier, speech itself might be regarded as an end result of progressive conventionalization.

Speech carries other advantages unrelated to conventionalization per se. While gesturing can be physically exhausting, the physiological costs of speech are so low as to be nearly unmeasurable (Russell et al. 1998). Speech adds almost nothing to the energetic cost of breathing, which we must do anyway to sustain life. Speech also frees the hands and body for other activities, such as manufacture and tool use, and allows us to verbally explain manufacturing techniques while at the same time demonstrating them physically. Indeed, it may have been this that led to a surge in creative artifacts that some described as the cultural revolution of the middle to late Pleistocene (Corballis 2004). Even so, we should not forget that manual gestures still play an important role in human language, and the signed languages of the deaf lose very little, if anything, in linguistic terms compared to speech.

With the increasing importance of the mouth, gestures may have begun to retreat inside the mouth, involving the tongue, lips, and eventually the larynx. Such gestures are largely invisible, and recovered through the addition of sound, which is modulated by internal gestures. In short, speech is facial gesture, half swallowed.[1]

Acknowledgments I owe thanks to many people, including (but not restricted to) Michael Arbib, Giovanni Berlucchi, Louis-Jean Boë, Tecumseh Fitch, Greg Hickok, William Hopkins, Nicholas Humphrey, James Hurford, Adam Kendon, Maurizio Gentilucci, Russell Gray, Susan Goldin-Meadow, David Leavens, Giacomo Rizzolatti, Thomas Suddendorf, and Virginia Volterra.

References

Annett, M. (2002). *Handedness and brain asymmetry: The right shift theory*. Park Drive: Psychology Press.

Arbib, M. A. (2005). From monkey-like action recognition to human language: An evolutionary framework for neurolinguistics. *Behavioral and Brain Sciences, 28*, 105–124.

Armstrong, D. F., Stokoe, W. C., & Wilcox, S. E. (1995). *Gesture and the nature of language*. Cambridge: Cambridge University Press.

Badzakova-Trajkov, G., Häberling, I. S., Roberts, R. P., & Corballis, M. C. (2010). Cerebral asymmetries: Complementary and independent processes. *PloS One, 5*, e9682.

Barney, A., Martelli, S., Serrurier, A., & Steele, J. (2012). Articulatory capacity of Neanderthals, a very recent and human-like fossil hominin. *Philosophical Transactions of the Royal Society of London. Series B, Biological Sciences, 367*, 88–102.

Biro, D., Sousa, C., & Matsuzawa, T. (2006). Ontogeny and cultural propagation of tool use by wild chimpanzees at Bossou, Guinea: Case studies in nut cracking and leaf folding. In T. Matsuzawa, M. Tomonaga, & M. Tanaka (Eds.), *Cognitive development in chimpanzees* (pp. 476–508). Tokyo: Springer.

Boë, L.-J., Heim, J.-L., Honda, K., Maeda, S., Badin, P., & Abry, C. (2007). The vocal tract of newborn humans and Neanderthals: Acoustic capabilities and consequences for the debate on the origin of language. A reply to Lieberman (2007a). *Journal of Phonetics, 35*, 564–581.

Boesch, C. (1991). Handedness in wild chimpanzees. *International Journal of Primatology, 12*, 541–558.

Boesch, C., Head, J., & Robbins, M. M. (2009). Complex tool sets for honey extraction among chimpanzees in Loango National Park, Gabon. *Journal of Human Evolution, 56*, 560–569.

Bogart, S. L., & Pruetz, J. D. (2008). Ecological context of savanna chimpanzee (Pan troglodytes Verus) termite fishing at Fongoli, Senegal. *American Journal of Primatology, 70*, 605–612.

Burling, R. (1999). Motivation, conventionalization, and arbitrariness in the origin of language. In B. J. King (Ed.), *The origins of language: What nonhuman primates can tell us* (pp. 307–350). Santa Fe, NM: School of American Research Press.

Calvert, G. A., & Campbell, R. (2003). Reading speech from still and moving faces: The neural substrates of visible speech. *Journal of Cognitive Neuroscience, 15*, 57–70.

Capirci, O., & Volterra, V. (2008). Gesture and speech: The emergence and development of a strong and changing partnership. *Gesture, 8*, 22–44.

Chomsky, N. (2007). Biolinguistic explorations: Design, development, evolution. *International Journal of Philosophical Studies, 15*, 1–21.

Chomsky, N. (2010). Some simple evo devo theses: How true might they be for language. In R. K. Larson, V. Déprez, & H. Yamakido (Eds.), *The evolution of human language* (pp. 45–62). Cambridge: Cambridge University Press.

[1] And with tongue partly in cheek.

Christiansen, M. H., & Kirby, S. (2003). Language evolution: The hardest problem in science? *Studies in the Evolution of Language, 3*, 1–15.

Corballis, M. C. (1991). *The lopsided ape: Evolution of the generative mind, Oxford University Press on Demand*. Oxford: Oxford University Press.

Corballis, M. C. (2002). *From hand to mouth: The origins of language*. Princeton, NJ: Princeton University Press.

Corballis, M. C. (2004). The origins of modernity: Was autonomous speech the critical factor? *Psychological Review, 111*, 543.

Corballis, M. C. (2009). The evolution of language. *Annals of the New York Academy of Sciences, 1156*, 19–43.

Corballis, M. C., Badzakova-Trajkov, G., & Häberling, I. S. (2012). Right hand, left brain: Genetic and evolutionary bases of cerebral asymmetries for language and manual action. *Wiley Interdisciplinary Reviews: Cognitive Science, 3*, 1–17.

Critchley, M. (1939). *The language of gesture*. London: Edward Arnold & Co.

Critchley, M. (1975). *Silent language*. Oxford: Butterworth-Heinemann.

Darwin, C. (1859). *On the origin of species by means of natural selection*. London: John Murray.

de Condillac É. B. (1971) *An essay on the origin of human knowledge: Being a supplement to Mr. Locke's essay on the human understanding [Essai sur l'origine des connaissances humaines. English]*. Cambridge University Press: Cambridge. Scholars' Facsimiles & Reprints.

Donald, M. (1991). *Origins of the modern mind: Three stages in the evolution of culture and cognition*. Cambridge, MA: Harvard University Press.

Dunbar, R. (1998). *Grooming, gossip, and the evolution of language*. Cambridge, MA: Harvard University Press.

Egnor, S. R., & Hauser, M. D. (2004). A paradox in the evolution of primate vocal learning. *Trends in Neurosciences, 27*, 649–654.

Emmorey, K. (2001). *Language, cognition, and the brain: Insights from sign language research*. Mahwah, NJ: Lawrence Erlbaum Associates.

Enard, W., Przeworski, M., Fisher, S. E., Lai, C. S. L., Wiebe, V., et al. (2002). Molecular evolution of FOXP2, a gene involved in speech and language. *Nature, 418*, 869–872.

Evans, N. (2009). *Dying words: Endangered languages and what they have to tell us*. Oxford: Wiley-Blackwell.

Evans, N., & Levinson, S. C. (2009). The myth of language universals: Language diversity and its importance for cognitive science. *Behavioral and Brain Sciences, 32*, 429–448.

Everett, D. (2005). Cultural constraints on grammar and cognition in Piraha: Another look at the design features of human language. *Current Anthropology, 46*, 621–646.

Fonagy, P., & Target, M. (2007). The rooting of the mind in the body: New links between attachment theory and psychoanalytic thought. *Journal of the American Psychoanalytic Association, 55*, 411–456.

Foundas, A. L., Leonard, C. M., Gilmore, R. L., Fennell, E. B., & Heilman, K. M. (1996). Pars triangularis asymmetry and language dominance. *Proceedings of the National Academy of Sciences, 93*, 719–722.

Fowler, C. A., Shankweiler, D., & Studdert-Kennedy, M. (2016). "Perception of the speech code" revisited: Speech is alphabetic after all. *Psychological Review, 123*, 125.

Fujita, K. (2009). A prospect for evolutionary adequacy: Merge and the evolution and development of human language. *Biolinguistics, 3*, 128–153.

Galantucci, B., Fowler, C. A., & Turvey, M. T. (2006). The motor theory of speech perception reviewed. *Psychonomic Bulletin & Review, 13*, 361–377.

Gannon, P. J., Holloway, R. L., Broadfield, D. C., & Braun, A. R. (1998). Asymmetry of chimpanzee planum temporale: Humanlike pattern of Wernicke's brain language area homolog. *Science, 279*, 220–222.

Gardner, R. A., & Gardner, B. T. (1969). Teaching sign language to a chimpanzee. *Science, 165*, 664–672.

Goldin-Meadow, S., McNeill, D., & Singleton, J. (1996). Silence is liberating: Removing the handcuffs on grammatical expression in the manual modality. *Psychological Review, 103*, 34–54.
Goodall, J. (1986). *The chimpanzees of Gombe: Patterns of behavior*. Cambridge, MA: Belknap Press.
Gould, S. J. (1977). *Ontogeny and phylogeny*. Cambridge, MA: Harvard University Press.
Green, R. E., Krause, J., Briggs, A. W., Maricic, T., Stenzel, U., et al. (2010). A draft sequence of the Neandertal genome. *Science, 328*, 710–722.
Hayes, C. (1951). *The ape in our house*. London: Gollancz.
Hewes, G. W., Andrew, R. J., Carini, L., Choe, H., Allen Gardner, R., et al. (1973). Primate communication and the gestural origin of language [and comments and reply]. *Current Anthropology, 14*, 5–24.
Hickok, G. (2009). Eight problems for the mirror neuron theory of action understanding in monkeys and humans. *Journal of Cognitive Neuroscience, 21*, 1229–1243.
Hickok, G., Pickell, H., Klima, E., & Bellugi, U. (2009). Neural dissociation in the production of lexical versus classifier signs in ASL: Distinct patterns of hemispheric asymmetry. *Neuropsychologia, 47*, 382–387.
Hickok, G., & Poeppel, D. (2007). The cortical organization of speech processing. *Nature Reviews Neuroscience, 8*, 393–402.
Hobaiter, C., & Byrne, R. W. (2011). Serial gesturing by wild chimpanzees: Its nature and function for communication. *Animal Cognition, 14*, 827–838.
Hopkins, W. D. (1996). Chimpanzee handedness revisited: 55 years since finch (1941). *Psychonomic Bulletin & Review, 3*, 449–457.
Hopkins, W. D., & Leavens, D. A. (1998). Hand use and gestural communication in chimpanzees (Pan troglodytes). *Journal of Comparative Psychology, 112*, 95.
Hopkins, W. D., Marino, L., Rilling, J. K., & MacGregor, L. A. (1998). Planum temporale asymmetries in great apes as revealed by magnetic resonance imaging (MRI). *Neuroreport, 9*, 2913–2918.
Hopkins, W. D., & Nir, T. M. (2010). Planum temporale surface area and grey matter asymmetries in chimpanzees (Pan troglodytes): The effect of handedness and comparison with findings in humans. *Behavioural Brain Research, 208*, 436–443.
Hopkins, W. D., Taglialatela, J. P., & Leavens, D. A. (2007). Chimpanzees differentially produce novel vocalizations to capture the attention of a human. *Animal Behaviour, 73*, 281–286.
Humle, T., & Matsuzawa, T. (2009). Laterality in hand use across four tool-use behaviors among the wild chimpanzees of Bossou, Guinea, West Africa. *American Journal of Primatology, 71*, 40–48.
Iverson, J. M., & Goldin-Meadow, S. (2005). Gesture paves the way for language development. *Psychological Science, 16*, 367–371.
Jarvis, E. D. (2006). Selection for and against vocal learning in birds and mammals. *Ornithological Science, 5*, 5–14.
Johansson, S. (2013). The talking Neanderthals: What do fossils, genetics, and archeology say? *Biolinguistics, 7*, 35–74.
Jürgens, U. (2002). Neural pathways underlying vocal control. *Neuroscience & Biobehavioral Reviews, 26*, 235–258.
Kaminski, J., Call, J., & Fischer, J. (2004). Word learning in a domestic dog: Evidence for "fast mapping". *Science, 304*, 1682–1683.
Kellogg, W. N., & Kellogg, L. A. (1933). The ape and the child: A study of environmental influence upon early behavior. *JAMA, 101*(15), 1181.
Kelly, S. D., Özyürek, A., & Maris, E. (2009). Two sides of the same coin speech and gesture mutually interact to enhance comprehension. *Psychological Science, 21*(2), 260–267.
Kendon, A. (2004). *Gesture: Visible action as utterance*. Cambridge: Cambridge University Press.
Kohler, E., Keysers, C., Umilta, M. A., Fogassi, L., Gallese, V., & Rizzolatti, G. (2002). Hearing sounds, understanding actions: Action representation in mirror neurons. *Science, 297*, 846–848.
Krause, J., Lalueza-Fox, C., Orlando, L., Enard, W., Green, R. E., et al. (2007). The derived FOXP2 variant of modern humans was shared with Neandertals. *Current Biology, 17*, 1908–1912.

Ladygina-Kohts, N. N., de Waal, F., & Wekker, B. T. (2002). *Infant chimpanzee and human child: A classic 1935 comparative study of ape emotions and intelligence*. Oxford: Oxford University Press.

Lieberman, P. (2007). The evolution of human speech: Its anatomical and neural bases. *Current Anthropology, 48*, 39–66.

Lonsdorf, E. V., & Hopkins, W. D. (2005). Wild chimpanzees show population-level handedness for tool use. *Proceedings of the National Academy of Sciences of the United States of America, 102*, 12634–12638.

Marean, C. W. (2010). Pinnacle point cave 13B (Western Cape Province, South Africa) in context: The cape floral kingdom, shellfish, and modern human origins. *Journal of Human Evolution, 59*, 425–443.

Marean, C. W., Bar-Matthews, M., Bernatchez, J., Fisher, E., Goldberg, P., et al. (2007). Early human use of marine resources and pigment in South Africa during the Middle Pleistocene. *Nature, 449*, 905–908.

Marshall, A. J., Wrangham, R. W., & Arcadi, A. C. (1999). Does learning affect the structure of vocalizations in chimpanzees? *Animal Behaviour, 58*, 825–830.

McBrearty, S., & Brooks, A. S. (2000). The revolution that wasn't: A new interpretation of the origin of modern human behavior. *Journal of Human Evolution, 39*, 453–563.

McGurk, H., & MacDonald, J. (1976). Hearing lips and seeing voices. *Nature, 264*, 746–748.

McManus, C. (2002). *Right hand, left hand: The origins of asymmetry in brains, bodies, atoms and cultures*. London: Weidenfeld & Nicolson.

McNeill, D. (1985). So you think gestures are nonverbal? *Psychological Review, 92*, 350.

Mellars, P. (2005). The impossible coincidence. A single-species model for the origins of modern human behavior in Europe evolutionary. *Anthropology: Issues, News, and Reviews, 14*, 12–27.

Mestres-Missé, A., Turner, R., & Friederici, A. D. (2012). An anterior–posterior gradient of cognitive control within the dorsomedial striatum. *NeuroImage, 62*, 41–47.

Meyer, M., Kircher, M., Gansauge, M.-T., Li, H., Racimo, F., et al. (2012). A high-coverage genome sequence from an archaic Denisovan individual. *Science, 338*, 222–226.

Mithun, M. (1999). *The languages of native North America*. Cambridge: Cambridge University Press.

Mukamel, R., Ekstrom, A. D., Kaplan, J., Iacoboni, M., & Fried, I. (2010). Single-neuron responses in humans during execution and observation of actions. *Current Biology, 20*, 750–756.

Patterson, F. G. P., & Gordon, W. (2001). Twenty-seven years of project Koko and Michael. In B. M. F. Galdikas, N. E. Briggs, L. K. Sheeran, G. L. Shapiro, & J. Goodall (Eds.), *All apes great and small: African apes* (Vol. 1, pp. 165–176). New York: Springer US.

Petkov, C. I., & Jarvis, E. (2012). Birds, primates, and spoken language origins: Behavioral phenotypes and neurobiological substrates. *Frontiers in Evolutionary Neuroscience, 4*, 12.

Pika, S., & Mitani, J. C. (2009). The directed scratch: Evidence for a referential gesture in chimpanzees. *The Prehistory of Language, 1*, 166–181.

Pilley, J. W., & Reid, A. K. (2011). Border collie comprehends object names as verbal referents. *Behavioural Processes, 86*, 184–195.

Pinker, S., & Bloom, P. (1990). Natural language and natural selection. *Behavioral and Brain Sciences, 13*, 707–727.

Pollick, A. S., & De Waal, F. B. (2007). Ape gestures and language evolution. *Proceedings of the National Academy of Sciences, 104*, 8184–8189.

Premack, D. (2007). Human and animal cognition: Continuity and discontinuity. *Proceedings of the National Academy of Sciences, 104*, 13861–13867.

Pruetz, J. D., & Bertolani, P. (2007). Savanna chimpanzees, Pan troglodytes Verus, hunt with tools. *Current Biology, 17*, 412–417.

Ptak, S. E., Enard, W., Wiebe, V., Hellmann, I., Krause, J., Lachmann, M., et al. (2009). Linkage disequilibrium extends across putative selected sites in FOXP2. *Molecular Biology and Evolution, 26*, 2181–2184.

Quintilian. (1920). *Institutio Oratoria XI: Loeb classical library edition.* (J. S. Watson, Trans.). Cambridge, MA: Harvard University Press.
Raffaele, P. (2006). Speaking bonobo. *Smithsonian Magazine, 37*, 74.
Reich, D., Green, R. E., Kircher, M., Krause, J., Patterson, N., et al. (2010). Genetic history of an archaic hominin group from Denisova cave in Siberia. *Nature, 468*, 1053–1060.
Rizzolatti, G., Camarda, R., Fogassi, L., Gentilucci, M., Luppino, G., & Matelli, M. (1988). Functional organization of inferior area 6 in the macaque monkey. *Experimental Brain Research, 71*, 491–507.
Rizzolatti, G., & Craighero, L. (2004). The mirror-neuron system. *Annual Review of Neuroscience, 27*, 169–192.
Rizzolatti, G., & Sinigaglia, C. (2008). *Mirrors in the brain: How our minds share actions and emotions*. Oxford: Oxford University Press.
Rizzolatti, G., & Sinigaglia, C. (2010). The functional role of the parieto-frontal mirror circuit: Interpretations and misinterpretations. *Nature Reviews Neuroscience, 11*, 264–274.
Rousseau, J. J. (1782). *Essai sur l'origine des langues. Geneva. English edition.*: (J. H. Moran & C. Alexander, Trans. (1986). On the origin of language). Chicago: University of Chicago Press.
Russell, B. A., Cerny, F. J., & Stathopoulos, E. T. (1998). Effects of varied vocal intensity on ventilation and energy expenditure in women and men. *Journal of Speech, Language, and Hearing Research, 41*, 239–248.
Savage-Rumbaugh, E. S., Shanker, S., & Taylor, T. J. (1998). *Apes, language, and the human mind.* Oxford: Oxford University Press on Demand.
Shea, J. J. (2010). Homo sapiens Is as Homo sapiens was: Behavioral variability versus "behavioral modernity" in Paleolithic archaeology. *Current Anthropology, 52*, 1–35.
Slocombe, K. E., Kaller, T., Call, J., & Zuberbühler, K. (2010). Chimpanzees extract social information from agonistic screams. *PloS One, 5*, e11473.
Slocombe, K. E., & Zuberbühler, K. (2005). Functionally referential communication in a chimpanzee. *Current Biology, 15*, 1779–1784.
Stokoe, W. C. (2001). *Language in hand: Why sign came before speech.* Washington D.C.: Gallaudet University Press.
Stout, D., & Chaminade, T. (2012). Stone tools, language and the brain in human evolution. *Phil Trans R Soc B, 367*, 75–87.
Studdert-Kennedy, M. (2005). How did language go discrete? In M. Tallerman (Ed.), *Language origins: Perspectives on evolution* (pp. 48–67). Oxford: Oxford University Press.
Tattersall, I. (2012). *Masters of the planet: The search for our human origins.* London: Macmillan.
Tomasello, M. (2008). *Origins of human communication.* Cambridge, MA: MIT press.
Tooby, J., & DeVore, I. (1987). The reconstruction of hominid evolution through strategic modeling. In W. G. Kinzey (Ed.), *The evolution of human behavior: Primate models* (pp. 183–237). Albany, NY: SUNY Press.
Umilta, M. A., Kohler, E., Gallese, V., Fogassi, L., Fadiga, L., Keysers, C., et al. (2001). I know what you are doing: A neurophysiological study. *Neuron, 31*, 155–165.
van der Lely, H. K., & Pinker, S. (2014). The biological basis of language: Insight from developmental grammatical impairments. *Trends in Cognitive Sciences, 18*, 586–595.
Vingerhoets, G., Acke, F., Alderweireldt, A. S., Nys, J., Vandemaele, P., & Achten, E. (2012). Cerebral lateralization of praxis in right-and left-handedness: Same pattern, different strength. *Human Brain Mapping, 33*, 763–777.
Wada, J. A., Clarke, R., & Hamm, A. (1975). Cerebral hemispheric asymmetry in humans: Cortical speech zones in 100 adult and 100 infant brains. *Archives of Neurology, 32*, 239–246.
Watkins, K. E., Strafella, A. P., & Paus, T. (2003). Seeing and hearing speech excites the motor system involved in speech production. *Neuropsychologia, 41*, 989–994.
Wundt, W. (1900). *Die sprache* (Vol. 2). Leipzig: Enghelman.
Xu, J., Gannon, P. J., Emmorey, K., Smith, J. F., & Braun, A. R. (2009). Symbolic gestures and spoken language are processed by a common neural system. *Proceedings of the National Academy of Sciences, 106*, 20664–20669.

Index

M. Mody (ed.), *Neural Mechanisms of Language*, Innovations in Cognitive Neuroscience, DOI 10.1007/978-1-4939-7325-5

T

MIX
Papier aus verantwortungsvollen Quellen
Paper from responsible sources
FSC® C105338

If you have any concerns about our products,
you can contact us on
ProductSafety@springernature.com

In case Publisher is established outside the EU,
the EU authorized representative is:
Springer Nature Customer Service Center GmbH
Europaplatz 3, 69115 Heidelberg, Germany

Printed by Libri Plureos GmbH
in Hamburg, Germany